W9-BGL-489

WITHDRAWN

Document Imaging Technology

**How Automated
Solutions are
Revolutionizing
the Way Organizations
and People Work**

Edwin D. McDonell

ISBN 1-55738-336-7
Printed in the United States of America

TS

CTV/BN

1 2 3 4 5 6 7 8 9 0

For my wife, Katherine

"If only it weren't for the people," said Finnerty,
"always getting tangled up in the machinery.
If it weren't for them, earth would be an
engineer's paradise."

— from *Player Piano* by Kurt Vonnegut, Jr.

Contents

List of Figures

Acknowledgments

It is with much gratitude that I acknowledge those who partici-
pated directly or indirectly in the development of this book. My
research assistant, Kathleen Hawk, provided extensive interview notes from her conversations with imaging consultants and
users worldwide. Kathy is a free-lance writer. We became ac-
quainted several years ago when she called to interview me for
an article in *Bank Marketing*. We have since co-authored several
articles and while we have never met, I consider her a personal
friend. Our articles are usually exchanged via fax machine between St. Augustine, Florida, and Indianapolis, Indiana—a trib-
ute to the effectiveness of facsimile technology.

Production assistance was provided by Jill Bond, a former
editor with Prentice-Hall, who labored for several weeks organ-
izing my references and exhibits. Larry DesJardines, a senior
technology consultant with USA GROUP, served as my technical
editor and Richard Robbins, a network manager with USA
GROUP, prepared and organized the photographs. My production editor was Christopher Svare. Chris and I first worked

together when he was editor of Bank Administration Institute's *Bank Management* magazine, and it was a pleasure working with him again.

I would also like to express my appreciation for the advice, encouragement and patience of several managers at Probus Publishing. This is the second book that Jim McNeil, vice president of editorial, and Kevin Thornton, senior production manager, have helped me conceive and develop. A book is the sum total of many ideas, and the contribution of editors is rarely acknowledged.

Jim McNeil also deserves credit for selling the European rights to my first book, *Creating a Customer-Driven Retail Bank*. When I was mid-stream in developing this book on imaging, and in need of some rest and relaxation, Michael Lafferty of Lafferty Publications, Dublin, paid for my travel to London to help promote my book's European edition. Although Michael is a successful publisher, perhaps his greatest talent is as a gracious host who fosters friendships among bankers worldwide.

Sincere thanks are also due to my colleagues at USA GROUP, whom I learn from daily: Ted Pollack, Greg Clancy, Hamed Omar, Sharon Vincent and the many talented staff of our Office Automation Services Department. USA GROUP is an Indianapolis-based financial services company specializing in the origination, servicing and securitization of educational loans. It is a company where service quality is given high priority, as reflected by Roy Nicholson, USA GROUP's chairman, who exemplified the concept of "the leader as servant" long before it was popularized in *Harvard Business Review*. It is a pleasure and a privilege for me to contribute to this outstanding organization.

Finally, I want to express my appreciation to my wife, Kathy, for her support through the many months of research, writing and editing during preparation of this book. A published writer herself in the field of American history, Kathy under-stands—as only another writer can—the intense desire to craft words into a lasting contribution.

E.D.M., Indianapolis, Indiana

Introduction

The arrival of the "paperless office" has been heralded repeatedly over the last two decades. Until recently, however, technology only succeeded in creating an office with somewhat less paper. Consider the following:[1]

- More than 300 billion documents are on file in the United States, accumulating in an estimated 100 million file drawers.

- A typical executive stores the equivalent of five filing cabinets of paper. The annual cost to maintain the standard five-drawer file cabinet is estimated to range from $800 to $1,500.

- A typical executive also spends more than 150 hours annually searching for documents that were misplaced, misfiled or mislabeled. The average missing file costs $120 in lost productivity.

---

1. See references.

+ Ninety-five percent of the most crucial information used to run organizations is still available only in paper form. Only two percent is available through on-line data processing systems. The remaining three percent is stored as microfiche.

Document imaging provides the first opportunity to literally replace paper with computerized images. This technology instantly delivers a distinct image of the document to a user's computer screen. Unlike microfiche, a digitized image can be shared by several users simultaneously, can be integrated with other automated systems, and cannot be misplaced. When necessary, a hard copy can be reproduced using a laser printer.

The most advanced imaging technology, called multimedia, integrates both voice and data with a document's image. An electronic stylus and pad allow you to make notes on the document on the screen, while explaining (speaking into a telephone) what revisions should be made. Someone later viewing the document can hear your explanation and watch as your notes appear on the screen.

When science fiction author Arthur C. Clark wrote, "Any sufficiently advanced technology is indistinguishable from magic," he certainly could have been describing imaging technology. My goal with this book, however, is to demystify document imaging, an area of heavy investment and intense interest. In 1992, for example, an estimated $14.4 billion was invested on imaging technology worldwide. Annual sales are projected to be $30.7 billion by 1995.[2] Given the scope of these expenditures, the success of imaging is simply a business imperative.

While many organizations have experienced significant gains through document imaging, other efforts have faltered

---

2.   "The Corporate Electronic Imaging Market: A Macro Overview, 1991-95," (New York: *Imaging World*, 1991): 2.

through a misunderstanding or misapplication of this technology. And many promising pilot installations must still be expanded and integrated within organizations in order to achieve a full return on image investments.

When applied successfully, imaging technology fundamentally alters the way people work. This book is therefore intended for both the users of image technology and the professionals designing and implementing image systems. Imaging technology represents an historic convergence of data processing and the traditional paper workflows within an organization. As stated in *Forbes*, "This technology [imaging] and the capabilities that it brings to an organization will have the same pervasive impact on society that personal computing had in the 1980s, but on a much grander scale."[3]

It would be dangerous, however, to assume that imaging is a "turnkey" technology that will miraculously transform the workplace. Stated simply, imaging represents a double-or-nothing "bet" for your organization. Imaging will dramatically and irrevocably increase your fixed costs. Requirements include a local area network (LAN) supported by skilled technicians with elaborate diagnostic tools; powerful workstations with high-resolution monitors for each worker; expensive optical storage and retrieval systems (OSARs); and ongoing consulting fees for software customization, which may represent more than 50 percent of the initial cost of the software and hardware.

In 1982, service companies invested an average of $6,000 in capital for each white-collar worker. By 1992, that figure had doubled to $12,000.[4] And for each worker empowered by imaging, this technology will represent an estimated capital invest-

---

3. David Churbuck, "Computers' New Frontier," *Forbes* (November 26, 1990): 264.
4. Stephen S. Roach, "Services Under Siege—The Restructuring Imperative," *Harvard Business Review* (September-October 1991): 82-91.

ment of $20,000 or more. Given the size of these investments, it would be risky to consider imaging as simply a new way to automate old ways of doing business. In terms of risk/reward considerations, there is genuine cause for concern. According to Stephen Roach, a senior economist at Morgan Stanley & Company, there is virtually no correlation in America's service sector between investments in technology and increased productivity. As stated by Roach in *Harvard Business Review*:

> "Productivity benefits from information technology in open-ended office applications have been particularly elusive. For example, the seemingly attractive and very expensive concept of the fully networked office environment now rings increasingly hollow. Technology connects machines, but so far has done little to instill productive synergy among people. An interconnected office environment may facilitate the flow of electronic messages, but the creative, high-value applications are still lacking. And yet it is precisely those types of innovative, idea-driven breakthroughs that lie at the heart of America's long history of productivity enhancement."[5]

The service sector now holds more than 85 percent of America's installed base of information technology. The transition from a variable-cost to a fixed-cost structure, without any gains in productivity, has only made American business more rigid and less able to compete. Four of California's 10 largest banks—Union Bank, Bank of California, Sanwa and Sumitomo—are now foreign owned.

For many organizations, imaging will merely be a new "wonder drug" to feed their addiction to technology. For organizations willing to challenge comfortable assumptions, however,

---

5.   Ibid.

imaging can be part of a service strategy that helps ensure their survival. My goal in this book is not to help wire more walls, but to break down those walls. Only then will imaging deliver an appropriate return on investment. The stakes have never been higher.

Structure of the Book

There are three phases common to the development of all systems: planning, design and implementation.

This book, like document imaging, should serve as a bridge between technology and the end-user. The "systems-development life cycle" familiar to systems professionals will be followed, but questions basic to the design of a system will be posed from a layperson's perspective. Questions to be answered include:

♦ Why will document imaging enhance the performance of a given organization?

♦ What documents should be stored as digitized images?

♦ Where in the organization will these images be used?

♦ Who should be authorized to store, retrieve or modify the images?

♦ When will these images be needed?

♦ How will they be accessed and used?

If these questions are posed at the appropriate time, and answered with a knowledge of both the organization and the technology, the success of your project is virtually ensured. If the needs of the organization or the strengths and limitations of the technology are misunderstood, progress will only occur through trial and error.

Chapter 1, *Understanding Imaging*, presents an overview of the technology and a plan for internal education, to build the foundation for later success. The educational value of vendor demonstrations and industry contacts are discussed, along with ways to maximize the benefit of time spent on these activities. Common barriers to a successful project are identified, and will be revisited throughout the book.

Chapter 2, *The Organization and Imaging*, introduces the concept of imaging as a strategic technology, and describes ways to link imaging to organizational goals. An approach to identifying and ranking candidate areas for imaging is presented, to maximize benefits while minimizing risks. The concept of work re-engineering is described (along with examples of how it is performed) and the role of the imaging task force is defined, along with suggestions on who should be represented on this task force.

Chapter 3, *Communicating with Vendors*, describes the development of a Request for Proposal, a standardized format to convey document processing needs to potential vendor candidates. A quantitative methodology is presented for evaluating and ranking vendor solutions, based on functionality and cost. An example template for a Request for Proposal is included, along with references to help you locate suitable vendor solutions.

The quantification of costs and benefits is discussed in Chapter 4, *Building a Business Case*. All potential costs are identified, including intangible costs such as disruption to the organization. Techniques for quantifying strategic benefits such as improved customer service are discussed, and example quanti-fications are included. Related issues—such as the cost of not pursuing imaging—are presented, and less complex, more intui-tive means of cost-justifying imaging are described. Finally, a "net-tangible-cost model" is included to help you weigh the tangible, or hard-dollar, costs of imaging against the potential strategic gains.

Chapter 5, *Vendor Evaluation*, explains the application of the benefit/cost methodology that was first introduced in Chapter 3. Techniques for efficiently validating claims made in vendor proposals are presented. A user survey to help identify vendor-support problems is included, along with criteria to help elimi-nate unsuitable vendors from further evaluation. The use of the benefit/cost methodology as a tool in contract negotiation is also explained.

Chapter 6, *System Design and Development*, provides an over-view of the logical and physical design phases that follow once an appropriate software and hardware platform has been se-lected. Technical considerations, such as the implications of client/server computing, are discussed. Issues relating to inte-gration of imaging with existing applications and other storage media are presented. Operational considerations in the design process, such as forms design and legal issues, are also identified.

Chapter 7, *Installation and Training*, addresses the release of the system, including testing, site preparation, and the backfile conversion of documents. Test issues and responsibilities (for both purchaser and vendor) are identified. Site requirements, including LAN infrastructure issues, are discussed. A technique for determining backfile conversion requirements is included, along with a checklist of potential conversion problems to avoid.

End-user training issues relating both to procedural changes and the system are explored.

Chapter 8, *Evaluation and Fine-Tuning*, examines the entire organization as a "system," following the "S diagram" used by McKinsey & Co. The potential impact of imaging technology on the organization's structure, staffing, skill-levels and management style is examined. Case studies from a variety of industries are presented, exploring issues of organizational change resulting from automation. The need to create a "learning organization" is discussed as a way to enhance and expand the imaging application over time in support of your organization.

In his book, *Technology, Management and Society*, Peter F. Drucker wrote: "What we lack, primarily, are the large concepts that will enable us to use these machines." This statement, made more than 20 years ago concerning the mainframe computer, is true today for imaging systems. My goal in *Document Imaging Technology* is to place this technology in context, so rather than merely being a tool, it can be a source of practical business solutions.

Introduction References

1. Behera, Bailochan. "Optical Storage Performance Modeling and Evaluation." *Optical Information Systems*, September-October 1990: 275.

   Elms, Theresa. "Image Processing: A New Vision for Information Systems." *ComputerWorld*, December 12, 1989: 29.

   "Imaging White Paper: Results from the Working Group." *Optical Information Systems*, May-June 1990: 140.

O'Connor, Mary Ann. "Applications of Optical Technology: Data Capture and Media Conversion." Document Image Automation, March-April 1991: 98.

Ryan, Alan J. "Imaging Technology Offers More Than Just Paper Gains." ComputerWorld, June 25, 1990: 14.

1

Understanding Imaging

"We are so accustomed to paper as a medium in our social and business exchanges, we often overlook the situations where paper has always performed poorly."
—Robert Zeek, Pfizer Corporation [1]

The greatest challenge in the application of imaging technology may be the eradication of what is called the "paper-based mindset." Every aspect of our work environment, from the size and arrangement of our desk drawers to the layout of our offices, reflects our reliance on paper. The daily lives of millions are designed to accommodate the limitations of paper and pencil.

Every new technology, of course, has faced a similar challenge. In the late 1950s, IBM, General Electric, RCA and 18 other major corporations declined patent rights to the photocopier. The prototype for the first commercially produced copier was

---

1. Robert Zeek, "Digital Document Image Automation: When Paper Fails," *Optical Information Systems*, March-April 1988: 50.

built with parts salvaged from scrap yards because the Haloid Company (later Xerox Corporation) lacked investors. Few could envision the practical benefits of this technology when an inexpensive alternative—carbon paper—already existed. A $10,000 investment in Xerox in 1960 was worth more than $1 million by 1972.[2]

A research project in 1990 co-sponsored by Wang Laboratories and the Nolan Norton Institute identified the following three critical obstacles to successful document imaging:[3]

- ♦ lack of knowledge concerning imaging
- ♦ organizational barriers
- ♦ technological barriers

Issues of organizational barriers and technological barriers are addressed in subsequent chapters. The foundation for success begins with knowledge. Writing is sometimes called the first "information technology." As with all successful technologies, the written word has become so pervasive we are hardly aware of its existence. Because the technology of paper and pencil has become essentially "invisible," we must force ourselves to examine both its strengths and weaknesses.

As noted by one user of document imaging, Robert Zeek, we can do the following with a piece of paper:[4]

- ♦ read it
- ♦ sign it
- ♦ copy and distribute it

2. David Owen, "Copies in Seconds," *The Atlantic*, February 1986: 67-69.
3. "Imaging White Paper: Results from the Working Group," *Optical Information Systems*, May-June 1990: 146.
4. Robert Zeek, "Digital Document Image Automation: Keeping It Simple," *Optical Information Systems*, May-June 1988: 99.

- take it home, to meetings, or on trips
- make notes on it or alter it
- shred it
- store it in warehouses

However, we cannot do the following with a piece of paper:

- find it reliably and quickly
- store it cheaply, and still find it quickly
- redo it cheaply
- know its status
- have more than one person work with it at a time
- combine audio and video comments
- index every word
- index it several ways, without copying it
- integrate it with existing applications

As Robert Zeek has remarked, the goal of an electronic image is to be "like paper, only better." This common-sense yardstick of value can help guide the layperson through a labyrinth of technical issues. In reviewing the technical data presented here or in vendor demonstrations, the user should appreciate that he or she is the ultimate authority on whether a technical feature is "like paper, only better."

Imaging Defined

As defined by the Association for Information and Image Management (AIIM), "Imaging is the ability to capture, store, retrieve, display, process and manage business information in digital form."[5] Each aspect of this definition is discussed in the following sections. The components of an imaging system are illustrated in Figure 1.1.

Figure 1.1 Components of Document Imaging System

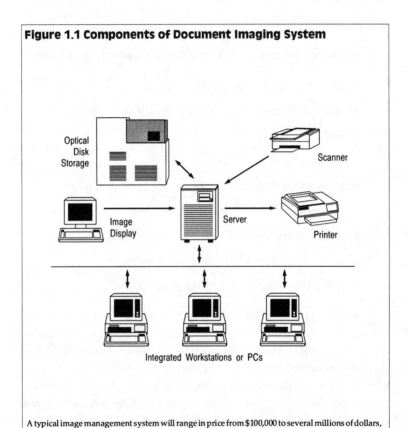

A typical image management system will range in price from $100,000 to several millions of dollars, depending upon configuration and volumes. The scanner digitizes documents and enters them into the system. The image display is used for indexing the documents for future retrieval. Compressed and indexed documents are stored on optical disks in a "jukebox." Users access the documents on PC workstations. Servers control the storing and printing of the documents.

---

5.   James E. Breuer, "Managing the Growth; AIIM Strategic Plan: Mission Statement, Goals, and Objectives," *Inform*, November-December, 1990: 32-34.

Image Capture

A newspaper photograph, closely examined, consists of thousands of black dots, arranged to duplicate the image of the original photograph. A *scanner* captured the image of the photograph and translated it into digitized data that could be stored electronically, transmitted and reproduced. Despite recent advances, this is not a new technology. The first electronic transmission of a news photo occurred on June 11, 1922, from Rome to New York. The photograph, of Pope Pius XI, appeared the same day in the *New York World*.

Document imaging employs a similar technology. The laser in a scanner reflects light onto a printed page, illuminating its pattern of light and dark. To record this image, a logical grid called a *raster pattern* or *bit map* is created to establish the location of the black dots. The actual recording is done with a medium that becomes electrically charged when exposed to light. Photocopiers and fax machines employ a similar technology when they duplicate or record images.

The design and sophistication of the scanner varies based on the nature of the graphic image to be recorded. Graphic images can be categorized as follows:

- ◆ Line art—white and black text or illustrations. This category includes black and white forms without gray or colored shading.

- ◆ Continuous tones that include shades of gray, as in a black and white photograph. This category includes forms that contain gray backgrounds.

- ◆ Color, as in a color photograph. This category includes forms that contain color printing or shading.

Your choice of a scanner will be based on the images to be captured. Images with more detail or shading require scanners with higher resolution (that is, *dots-per-inch*, or DPI). To ensure a crisp image is captured by the scanner, you may also need to redesign some of your forms when you adopt imaging technology. With the exception of publishing and design firms, color scanners are not frequently employed in business because of their high cost.

Universal II Scanner

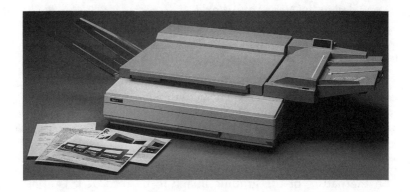

Corporate Design Association Inc.

Storage

When scanned, images are compressed to reduce data storage requirements. In simple terms, compression removes the white spaces from a document. Only the black and gray markings are stored. Compression may be performed using hardware (called a *compression board*) or software. Even with compression, the storage space required for images is still significant. A page of electronic text requires perhaps 4,000 bytes (4 KB) of memory, while the same page stored as an image demands approximately

40,000 bytes (40 KB)—10 times the space. Intensive research is being focused on more efficient algorithms for image compression.

Storage of large volumes of document images became economically feasible in the 1980s with the advent of optical storage technology. Magnetic storage (which includes fixed-disk, diskettes and tape) had been the *de facto* standard for data storage since the 1960s. Fixed-disk is typically called hard disk when used with personal computers or PCs, and DASD (Direct Access Storage Devices) with mainframe systems. Whatever it is called, fixed-disk has the advantage of a fast retrieval time—within several seconds. By contrast, magnetic tape and diskettes have an advantage of lower storage cost, because they can be stored in a library and mounted only as required. Traditionally, fixed magnetic disks are used for data required on-line, while tape or diskettes are used when archiving or exchanging data (for example, when a "back up" is performed for data security). PC users are familiar with the 3.5-inch and 5.25-inch formats for diskettes, and maintain a library of these for data storage.

From a functional viewpoint, optical storage represents an effective compromise between the advantages of fixed magnetic disk and tape. For large volumes of data, optical storage is less expensive than fixed magnetic disk, while still offering faster retrieval time than magnetic tape. The robotic technology of optical "jukeboxes" has proven generally reliable. Mechanical devices to mount tapes have never gained wide acceptance. A summary of storage media cost and access speed is presented in Figure 1.2.

Optical storage uses a high-powered laser beam to write data on a disk or tape by burning microscopic holes, called *pits*, into their surface. The data is then read by a low-powered laser beam, in much the same way that old paper tape and card readers interpreted information—by using reflected light to detect the presence or absence of a hole.

Figure 1.2 Comparison of Storage Media Cost and Performance

Media	Access Time	Cost/megabyte ($)
Computer memory	1 microsecond	1000.00
Hard disk	15 milliseconds	10.00
5.25 high density floppy	seconds	1.25
Magnetic tape (6250 bpi)	seconds	0.067
COM microfiche	seconds-minutes	0.75
5.25-inch optical disk*	milliseconds	0.166
12-inch optical disk*	145 milliseconds	0.055
Optical jukebox	5 seconds	0.15
Optical library	minutes/hours	0.02

Source: Bailochan Behera. "Jukebox Cost Performance Modeling." *Document Image Automation,* May-June 1991: 151.

Three types of optical storage are in commercial use today: *prerecorded, writable* and *rewritable.* Prerecorded optical storage, called CD-ROM (Compact-Disk, Read-Only Memory), is used for the distribution of large informational databases. One CD-ROM disk can accommodate an entire encyclopedia, and the cost of a CD-ROM reader is less than $600. Writable optical storage is called WORM (Write-Once, Read Many). WORM offers one-time writing of data, and unlimited reading (retrieval) of this stored data. Once recorded, the original data cannot be modified, but it can be updated by writing a new file elsewhere on disk, and linking the two files via a software pointer. When a read operation occurs, the new updated file is retrieved, although the original file still exists. WORM provides a tamper-proof audit trail, and is thus especially suited to mass storage of archival data.

Optical Disks

Corporate Design Association Inc.

Rewritable (sometimes called *erasable*) is the most recent innovation in optical storage. Rewritable optical storage provides the same read-and-write capabilities as traditional magnetic media, with the added benefit of mass storage capacity. The most widely used rewritable optical technology is called *magneto-optical*, or M-O. M-O is an optically assisted form of magnetic recording that uses a laser to heat selected spots on a recording surface. These spots, once heated, are susceptible to magnetism and can be used to record data. When the heated spot on the disk returns to normal temperature, it is again resistant to magnetism, making the recorded data more stable than with ordinary magnetic media. Figure 1.3 presents a functional comparison of storage media.

While optical disk storage made imaging feasible in the 1980s, it should not be confused with imaging. Contrary to popular belief, image processing is not synonymous with optical storage. Rather, optical disk is only one of several mass-storage technologies. Other approaches to mass storage, such as digital audiotape (DAT), optical tape and holographic memory, may emerge as competitors to optical disk in the 1990s.

Display

A document-imaging workstation is typically a PC augmented by a high-resolution monitor and an image-compression board. The computing power required to manipulate an image (magnify, overlay or annotate it) usually dictates a *client/server* environment. Client/server systems separate the storage of data and

Figure 1.3 Media Application Comparison

Media Type	Application Requirements								
	Storage			Distribution			Retrieval		
	Storage Time/Cost Capacity	Production Automation	Longevity	Duplication Time and Cost	Distribution Time and Cost	Wide Distribution Address	Retrieval Time	Retrieval Cost	Retrieval Ease
DASD	Low	High	Low	Medium	High	High	High	Medium	High
WORM	Medium	Medium	Medium	Medium	High	Medium	Medium	Medium	High
Erasable OD	Medium	High	Low	Medium	High	High	Medium	Medium	High
Tape/Tape Library	High	Medium	Low	Medium	High	Medium	Medium	Medium	Medium
CD-Writable	Medium	Low	Low	Medium	Medium	Medium	Medium	Medium	Medium
COM Fiche	High	Low	High	High	Medium	High	Low	Low	Low
COM Film	High	Low	High	Medium	Medium	High	Low	Low	Low

Relative Performance — High ● Medium ◐ Low ○

Source: Kodak Data Processing Products. 1991.

images from routine processing tasks. A mainframe or minicomputer *server* controls the centralized optical storage and manages communication to several *client* workstations used to view or process the image.

When an image is requested by a client workstation, it is retrieved from centralized optical storage by the server and transmitted via local area network (or LAN) to the workstation. It is then decompressed by the client workstation and stored locally on optical storage, or the PC's magnetic disk or internal memory (called *random access memory*, or RAM). It can then be viewed and manipulated locally, without further demands on the server. In an efficient client/server environment, a single server may support dozens of workstation clients. The result is the best of both worlds—PC performance combined with mainframe data access.

Unless a PC workstation includes a small optical storage peripheral, its primary distinction is its high-resolution monitor. The resolution or visual clarity of a monitor is stated in *dots-per-inch*, or DPI (also sometimes referred to as *picture-elements-per-inch* or *pixels*). Most PCs employ low-resolution monitors displaying 100 DPI. Low-resolution monitors (such as a VGA monitor) may be appropriate for the casual user who works primarily with text and only occasionally with images. To prevent eyestrain for frequent imaging users, however, the current trend is toward resolutions from 150 DPI to 300 DPI. Higher resolution displays (up to 400 DPI) are available but are not typically required in business applications. A 19-inch diagonal screen is ideal to display larger documents, but 17-inch or 14-inch monitors may be suitable for some applications.

Retrieval

Most organizations that use imaging will need the capability to produce, on demand, a hard copy of the electronic image. Laser-page printers, ubiquitous in most offices, can reproduce images

Document Imaging Workstation

as well as text. A printer-controller on the server converts the image data back into a grid pattern of dots (again, called a bit map), and transmits this datastream to the printer. The printer then "writes" the image onto a drum or belt that is electrically charged in a pattern that mirrors the original scanned image. This drum or belt then attracts positively charged toner particles to reproduce the image. Laser printers, fax machines and copiers all employ variations on a similar technology, which is called a *charge-coupled device*, or CCD.

Types of laser printers are distinguished by document resolution and their throughput or speed. Document resolution (again, stated as dots-per-inch or DPI) should be sufficient to reproduce an exact or near replica of the original. 300 to 600 DPI is acceptable for most business applications. Effective reproduction of some documents, such as photographs, requires higher resolution. The speed of laser printers is stated in pages per minute (PPM). Throughput is sometimes used to define the capability of the laser to support continued, heavy-duty production. Rugged, high-capacity laser printers—such as those used to produce your phone bill or credit-card statement—are limited only by the tendency of paper to catch fire from the friction created at very high speeds.

Laser Printer

Processing and Management

The first imaging systems introduced commercially were individual workstations. A single operator scanned, indexed, stored and retrieved the documents. These systems are often called *electronic file cabinets*. They are typically purchased by records managers and are appropriate for archival storage—that is, storage of documents that are relatively inactive. Many documents, however, are active; they are reviewed, processed and routed from desk to desk. They may be archived, but only after a business event is completed—after the loan is granted, the purchase is authorized or the claim is honored. The most advanced imaging solutions automate the business event. Both the paper and the process are managed, through the system's program.

This is facilitated by *optical character recognition* (OCR), which enables the data on the scanned document to be read and interpreted. A rudimentary form of OCR was used in grading the Iowa Standardized Tests, familiar to a generation of students, where multiple-choice answers were marked in pencil.

Through OCR, a bank check without a signature may be automatically flagged for review. A loan request exceeding a branch manager's lending authority may be routed automatically to the main office. Insurance claims may be prioritized automatically, based on the dollar amount of the claim or the date submitted. The review, prioritization and processing of the documents is supported by the system. While traditional office automation supports a *task*, such as word processing, imaging systems support a *process*.

A layperson unacquainted with imaging should take comfort in knowing that he or she is, in fact, probably familiar with all the technologies included in an imaging system. Even advanced multimedia applications simply refine and integrate well-established technologies used in photocopiers, fax machines, laser printers, video, voice mail and relational databases. It is therefore not surprising that imaging has been called a "a hearty technological soup."

The User and Imaging

A description of technical components in an imaging system, however elaborate, cannot convey the scope of user capabilities. To understand potential benefits, it is necessary to consider the flexibility imaging provides. Once a document's image is scanned, the document is freed from the constraints associated with paper-based filing and retrieval systems.

Traditional document processing is not unlike an assembly line. With a loan application, for example, a number of manual steps occur in a predetermined sequence: forms are reviewed for signatures, data such as an applicant's income is copied from the form to help make the credit decision, a credit bureau inquiry is made, and the application is ultimately passed from clerical staff to a loan officer for approval. Dozens of steps may occur in the process, and the documents may be handled by four or five staff. Although a few steps may be automated (such as credit-scoring) the effort is often manual. Figure 1.4 presents a paper-based process.

How would this process differ with imaging? First of all, simultaneous activity on the document can occur. While one person is viewing the document to perform employment verification, another can perform a credit inquiry, and yet another can

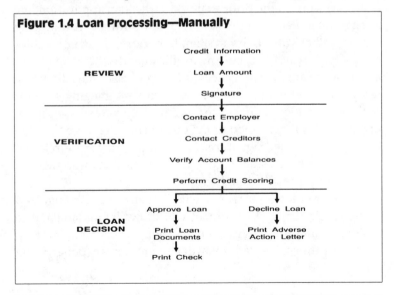

Figure 1.4 Loan Processing—Manually

perform the credit-scoring. Rather than a paper "assembly line" where activities occur in sequence, these activities can occur simultaneously.

What else does imaging offer? With paper-based forms, multiple indexing of documents is typically infeasible. When filing paperwork, one index method is chosen, such as customer name, account number or date processed. Once chosen, the index method is rarely altered, and no other means of accessing the document is possible (unless, of course, photocopies are made and stored). With imaged documents, there is no limit to the number of indexing methods. The same document can be stored and accessed later by date received, by customer or by department. This additional indexing adds little to the cost of electronic storage (while filing a photocopy would clearly double your storage costs). And the same document can be quickly retrieved by dozens of staff, throughout an office or throughout the country.

What else does imaging offer? Because the document is in electronic form, it can be integrated with the existing data processing systems. In the prior example of the loan application, the imaged documents might be integrated with a credit-scoring system and credit-inquiry system, to perform these tasks automatically. Research is occurring to allow handwritten data (such as the dollar amount of a check) to be read electronically using recognition software. Logic within document imaging systems can also provide *intelligent routing* of documents based upon the information contained in the document. A loan application lacking a customer signature, for example, might be rejected from further processing. Figure 1.5 presents a loan-origination process supported by imaging.

Organizations successful with imaging have recognized that re-engineering of document workflows is fundamental to achieving full productivity gains. Although document imaging can reduce the cost of traditional document storage (paper, filing

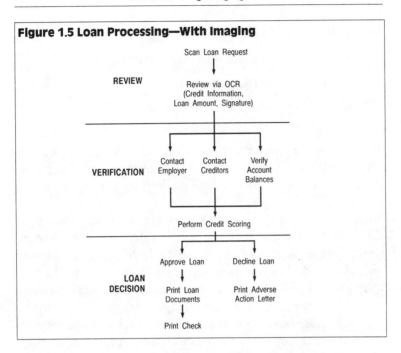

Figure 1.5 Loan Processing—With Imaging

space and clerical support), the greatest benefits are achieved through more efficient workflows and more effective decision-making.

Demonstrations

To help place the concepts presented in context, educational reading must be supplemented with hands-on experience. As with other visual technologies, such as graphical user interfaces (GUIs), it is impossible to fully appreciate the look and feel of the technology from a text description.

No marketing or technical literature can match in value a well-planned vendor demonstration. The ideal format should include an executive overview of the following:

- imaging as a strategic technology

- imaging applications in your industry

- example cost/benefit studies

- imaging project management issues

For smaller organizations, demonstrations can be accomplished through public, vendor-sponsored seminars. Larger organizations may find vendors willing to present in-house seminars. These seminars often include demonstrations of the more advanced imaging technologies, such as multimedia applications integrating voice, data, image and video. Vendor-sponsored public seminars may also include leading industry consultants. A directory of speakers on document imaging, organized by topic and geographic locations, is available from the Association for Information and Image Management (AIIM).[6]

AIIM also holds an annual conference, which includes presentations by organizations and consultants experienced in imaging technology. The AIIM conference is well-attended by vendors with demonstration booths, and can serve as an effective starting point for managers investigating this technology. Discussions with industry peers that have implemented imaging is another valuable source of information. As an AIIM research study notes, "Imaging is not a spectator sport management can observe from the sidelines."[7]

While all managers may benefit from an overview demonstration, first-level managers (that is, managers directly supervising paperflows) should receive demonstrations tailored to their own departments. Once overview seminars have whetted your organization's appetite for learning, in-house demos can be

---

6. *Speakers Directory*, Association for Information and Image Management, (Silver Spring, Md.: AIIM, 1993).
7. Association for Information and Image Management, "The Myths and Realities of Imaging," *Inform*, September 1992: A4-A8.

scheduled that present specific, relevant solutions for paper-handling functions, such as customer service, accounts receivable or claims processing. The vendor can perform a "walk-through" of your organization, gather relevant documents and prepare one or more demonstrations tailored to the needs of these departments.

These demos should also be attended by your information systems (IS) staff, especially those familiar with application systems that may later be integrated with image technology. If your organization has a separate records department, its manager might also attend. Additional attendees may include internal productivity consultants, customer-service managers, database administrators and audit staff. Some attendees will later serve on a task force to assist in implementing document imaging in your organization. The foundation for your future success can be established early, through education and participation.

Selecting a Vendor

Most demonstrations are initiated by a vendor. An imaging demonstration may be widely attended, and if the vendor is poorly prepared or if the vendor's system is flawed, these factors will influence attitudes throughout your organization for months or years.

First impressions are important. Because an organization's early perception of imaging technology is influenced by these demonstrations, an appropriate vendor should be invited to present a demonstration. In selecting a vendor to present a demonstration, an understanding of the industry's distribution channels can help (Figure 1.6). A distinction can be made between developers (those manufacturing or developing the systems) and marketers (those selling and installing the systems). Developers include:

Figure 1.6 Distribution Channels for Imaging Systems

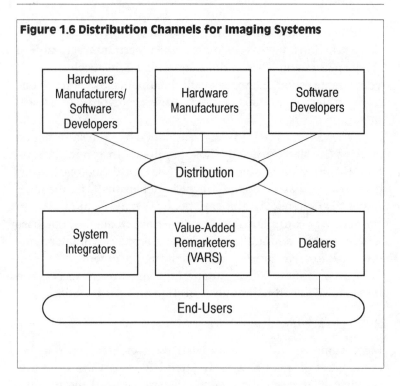

- ♦ hardware manufacturers/software developers
- ♦ software developers
- ♦ hardware manufacturers

Hardware Manufacturers/Software Developers

Virtually all major hardware vendors (including DEC, Hewlett-Packard, IBM, NCR and Unisys) also have imaging software. This imaging software may have been internally developed or it may be software they have acquired the right to market (known as third-party software). Hardware and software vendors can be further categorized as those supporting *open systems* (meaning adherence to industry standard technologies, such as the UNIX operating system), and those offering solutions based on non-standard architectures (such as a proprietary operating system).

Software Developers As noted previously, the first imaging systems introduced were electronic filing cabinets. These were (and many still are) systems with interdependent, proprietary hardware and software. Both hardware and software were inseparable and specific to the vendor. However, a number of these vendors have now abandoned the sale of imaging hardware to market open systems imaging software. Open systems software can be installed in any hardware environment compatible with non-proprietary operating systems, such as DOS or UNIX. Plexus Corporation's XDP software, for example, can be installed on UNIX-based hardware sold by Data General, DEC, Hewlett-Packard, IBM, NCR or Unisys. Some software providers also occupy specialized market niches. Calera, for example, is well-known as a developer of optical character recognition (OCR) software.

Hardware Manufacturers Any processor manufacturer that supports the UNIX operating system, even if it has no proprietary software, could provide a platform for a UNIX-based imaging system. This category also includes the numerous vendors of specialized peripheral hardware: scanners, displays, printers, optical storage and retrieval systems (OSARs) and jukeboxes. Just as some electronic file cabinet vendors have abandoned the hardware market, others have abandoned software sales to focus solely on specialized hardware, such as optical disk jukeboxes. These specialized hardware vendors typically have a strong foundation in research and development. Some of the more advanced monitors and mass-storage devices are marketed by small, research-oriented manufacturers. Cygnet, for example, is a noted developer of optical jukeboxes or OSARs.

Developers may or may not market their own systems. Marketers of imaging systems include:

♦ value-added remarketers (or VARs)

♦ dealers

♦ systems integrators

Value-Added Remarketers (VARS) VARs are marketers that provide additional consulting services. With the exception of selecting software (since they are already committed to a specific solution), they may provide all the services of a systems integrator. At a minimum, VARs offer software customization, consulting and support services. Many consulting firms also serve as VARs for specific imaging systems. When engaging a consulting firm to assist in the selection of an imaging solution, it is essential to determine if they have any existing VAR relationships that could prejudice their decision-making ability.

Dealers Dealers provide few services, other than the delivery and set-up of the imaging equipment and on-going hardware maintenance services. Many major hardware manufacturers, such as Hewlett-Packard, are classed as dealers, but have extensive marketing alliances with consulting firms that serve as their VARs.

Systems Integrators Systems integrators are not marketers in the strict sense of the term, although they will advise in a purchasing decision. Integrators are consultants specializing in the selection, integration and customization of imaging systems. Some dealers or VARs present themselves as integrators but, in the strictest sense, an integrator is independent of specific technical solutions. Integrators may be employed when complex multimedia applications are desired (that is, systems combining voice, data, image or video). They may also link diverse vendor solutions (for example, integrating a departmental imaging system with a corporate mainframe). Integrators will play an increasingly important role as imaging systems serve to bridge what consultant Thornton May has called the "islands of technology" within an organization.

As noted by another consultant, Robert Kalthoff, the role of an imaging integrator is broader than the traditional role of an EDP (electronic data processing) integrator. In addition to the technical integration of diverse software and hardware solutions, the imaging integrator may "provide everything from needs analysis, systems design, application software and system services, to installation, booting-up, training, and on-going upgrades."[8] As discussed later, the roles of vendors and integrators are often blurred.

Arranging a Demonstration

By default, the first demonstration or "demo" is often by an organization's current hardware vendor. These vendors may be limited to a specific proprietary solution. Many demos are also offered by electronic file cabinet vendors. While these systems are highly effective for their intended role, they present only a limited view of possible technologies. You may be wise to consider a demonstration by an integrator familiar with several technologies. Contact names for integrators are available through the membership directory of the Association for Information and Image Management (AIIM).[9]

---

8. Robert J. Kalthoff, "Large Scale Document Automation: The Systems Integration Issues," *Video and Optical Disk*, November-December 1985: 447-448.
9. Association for Information and Image Management, *Buying Guide and Membership Directory.* (Silver Spring, Md.: AIIM, 1993).

The most effective demos require an analysis by the vendor of your organization's paperflow. The concept of work redesign or re-engineering is discussed in detail in the following chapter. The vendor's analysis is similar in nature but more limited in scope. The vendor will identify general opportunities for restructuring your tasks through imaging. These ideas should then spark more detailed discussions, during and after the demo. Workflow consultants (who may also be integrators) often assist vendors in preparing demonstrations. These consultants may be employed by the vendor or may have a marketing relationship with the vendor. In either case, they can provide valuable insights. They may identify some benefits you can immediately achieve, prior to an investment in imaging.

To prepare for a demo, vendors typically request a *procedural walk-through*. This walk-through should chronologically follow the paper-handling process from beginning to end. The best guide for this walk-through is a first-level manager who directly supervises the tasks under review. They will either have the background to answer detailed questions, or can refer the consultant to someone who will. A skilled consultant can put staff at ease and develop the rapport needed later during implementation.

The vendor may ask you to gather and organize copies of your documents. These documents can then be scanned and incorporated into your demo to help you visualize the capabilities of their system. Copies of *completed* forms should be made, so that their use is clear to the vendor. Also, the forms gathered should correspond to the same transaction. Forms collected from a travel agency, for example, might relate to travel arrangements for a single customer. This enables the vendor to more quickly grasp the sequence of tasks and the relationship between the forms.

Structure of the Demonstration

A skilled vendor first presents an overview of the current paper-based process, followed by a demonstration of an alternate approach using imaging. He or she may provide workflow diagrams contrasting the paper-based and electronic processes, similar to Figures 1.4 and 1.5. A vendor may provide an estimate of potential savings or can often identify likely benefits, such as improved cash management or better customer service. The strongest vendors are aware of your industry's "critical success factors" and can directly relate an investment in imaging to your organization's stated goals.

Validating a vendor's understanding of corporate objectives is crucial in the vendor evaluation process (discussed in detail in Chapter 5) because many vendors bundle consulting and integration services with the hardware and software. If demonstrations are arranged with several vendors, these may provide preliminary insights into vendor capability. The goal of these discussions should be to provide a grasp of imaging capabilities from a *functional* viewpoint. The concept of multiple indexes for a single document is relevant, for example, while the particulars of a specific database are not.

Identify Candidates for Imaging

Initial demonstrations, combined perhaps with internal seminars, should enable your end-users to identify departments or functions that are candidates for imaging technology. This knowledge also allows end-users to employ their own creativity when revising paper-based tasks. Following the demonstrations, senior managers and first-level mangers should prepare a list of candidate areas for document imaging. From these candidates, your task force will later select the first application for imaging technology (for more information, refer to Chapter 2). When developing a candidate list, consider the following questions:

Is the paper generated external to your organization?

Many documents processed by a business are not produced internally—they arrive from customers and suppliers via mail or fax. When communication with a third-party is required, paperwork is often inevitable. Checks, credit applications, insurance claims and purchase orders are thus natural candidates for document imaging.

Is it impossible to capture the required information in a few keystrokes?

If the entire document needs to be viewed and not simply a few items of data, imaging may be valuable. An entire contract with signatures, for example, can be imaged by a law firm for later reference. If just a few items of information are needed before a document is destroyed, a database entry is more efficient than imaging. With a returned airline ticket, for example, an agent may simply enter an invoice number to credit a customer's account—and then shred the ticket to avoid further handling.

Can the paper flow be replaced with automation other than imaging?

When feasible, *electronic data capture* (EDC) and *electronic data interchange* (EDI) are preferable to paper or imaging. Account transfers performed at a bank's automated teller machine are an example of EDC—no manual processing is required. Direct-deposit payroll programs are a familiar example of EDI. For purchase orders, accounts payable and accounts receivable, EDI formats are rapidly being adopted to allow data exchange via magnetic tape or disk. EDI is very efficient when participants agree to specific technical standards. For example, Ford Motor Company recently demanded that all its suppliers conform to an industry-approved EDI format. When cooperation is not feasible, however, imaging can provide a solution.

Do the documents need to be retrieved by a number of individuals?

One of the earliest, widespread applications of imaging was for customer signature cards in banks and thrifts. A bank customer may seek to cash a check at any branch office—not just the one visited to open the account. Rather than ask a customer to sign multiple signature cards, many banks have installed imaging systems to display the signature on the teller terminal.

Does the paper require significant processing or handling?

Any paperwork handled in sequence by two or more individuals is a candidate for imaging, particularly when speed and control are important. Imaging systems can automatically route documents from desk-to-desk, and from office-to-office. They can also prioritize the processing of documents, and distribute workloads fairly between individuals or departments.

Must a number of documents be matched up before processing can occur?

When processing student-loan applications, for example, it is necessary to compare documents from students, parents and schools before authorizing a loan. Imaging can speed the collation of these documents. Images can also be integrated with an application system. When collecting bad debts, for example, a collector could view an aged receivables report on the same screen as an image of the bill-of-lading signed by the customer when the customer's order was delivered.

Do the documents contain information essential to the business or organizational mission?

When control of documents and immediate access to documents is mandatory, imaging is a solution. To prove compliance with the regulations of the Food and Drug Administration, for example, a pharmaceutical firm can employ imaging to index and archive documents concerning its monitoring of drug tests. If these documents were lost or destroyed, the cost to the firm might be incalculable, since retesting could require months or years, delaying introduction of an important drug.

Is speed a competitive advantage?

When speed-to-market for new products is a competitive strength or when prompt response to customer inquiries is essential, imaging is a strategic tool. For new product development, engineering departments can integrate computer-aided design (CAD) with document imaging and computer-aided manufacturing (CAM) to dramatically improve the speed-to-market for new products. Speed-to-market is closely correlated with market share, and market share is closely correlated with profitability.

Alternatives to Imaging

As noted, imaging may not be the first choice for an automated solution. Potential alternatives to imaging include the following:

♦ electronic data capture or interchange (EDC and EDI)

♦ micrographics

♦ voice mail

♦ facsimile

Electronic Data Capture

A cash withdrawal processed by an automated teller machine is an example of *electronic data capture*, or EDC. The debit to your account is passed electronically to your bank's deposit application system. The alternative, performing a paper-based transaction at your branch with a teller, can be improved through imaging but remains a second-best approach. The ideal solution is almost always electronic data capture, when it is feasible. Bar coding systems, widely used in retail stores, are another example of EDC technology.

A variant of electronic data capture is EDI, or *electronic data interchange*. As previously noted, EDI refers to the exchange of data, in a standardized electronic format, between organizations. If your monthly utility payment is debited from your checking account automatically, it is accomplished through EDI. If you use an automated teller machine belonging to another bank, the transaction is processed electronically through an Automated Clearing House or ACH transaction, which is a form of EDI.

Micrographics

Information systems departments turned to *computer-output microfilm* (COM) in the mid-'70s, when it became virtually impossible to print the volume of paper reports needed to satisfy user requirements. Computer-output microfilm (COM) now accounts for more than one-third of all computer output in the United States. An enhancement to COM, termed CAR (*computer-aided retrieval*), provides indexed access to this microfiche. In a manner somewhat similar to the optical disk jukebox, CAR systems physically retrieve and mount the requested fiche. CAR readers are similar to the usual microfiche readers, but do not require a user to handle the fiche.

Computer-output to laser disk (COLD) is somewhat analogous to COM/CAR, but with access to reports provided on-line through existing workstations. Advantages of COLD include faster, more convenient retrieval. The optical disk storage used for COLD, while less expensive than on-line magnetic storage, is more expensive than COM. COM possesses the significant advantages of low production cost and inexpensive, transportable access. COM can be distributed via mail to branch offices, for example, and accessed via a microfiche reader costing less than $500.

Although micrographics lack the glamour of document imaging, studies show microfiche has a long future. For example, developments in optical character recognition (OCR) now allow COM to be machine-readable as well as visually readable. The most likely scenario is that micrographic systems will become peripherals of document imaging systems for certain types of archival storage. Integrated COM/CAR/imaging systems could allow COM to be retrieved via a CAR index, scanned, digitized and transmitted to a user's screen. "Advances in other technologies, rather than threatening the viability of COM," according to David Bogue of Zytron Corporation, "are creating new roles for it."[10]

Voice Mail

While multimedia systems integrating voice and image are impressive in vendor demonstrations, voice enhancement is a relatively simple and inexpensive technology. If voice mail has been justified for inbound calls from third parties, the additional cost to integrate voice mail and imaging is inconsequential. Voice integration is an inexpensive add-on to an imaging system and can conveniently link a document to an audio message. Voice

---

10. David Bogue, "COM: Nothing That a Pseudonym Couldn't Fix," *Inform*, January 1991: 12-13, 29.

mail, however, can typically be cost-justified independent of any investment in imaging.

Facsimile

Facsimile, or fax, can be a peripheral in an imaging system or an alternative when imaging cannot be fully justified. A bank may choose to fax a customer signature card to an office when a check is cashed, rather than investing in high-resolution teller monitors to display the signature. An oil field equipment lessor in Texas selected fax machines as remote-input devices for a centralized imaging solution. Maintenance reports are faxed from regional offices (or even from hotels) to the firm's headquarters in Corpus Cristi. The imaging workstations used to index, store and retrieve the documents are only needed in their corporate office.

Enthusiasm and Realism

Eliminating the "paper-based mindset" is one goal of education, but the truly informed user will recognize both the strengths and weaknesses of various technologies—including the very enduring technology of paper and pencil. When reviewing your list of candidate areas for imaging, consider alternate solutions, perhaps developing an additional list of problems better addressed by other technologies.

Your enthusiasm for imaging should be tempered by the realization that not all paper can necessarily be eliminated. As imaging consultant Robert Kalthoff cautions: "Will all documents as we know them today disappear into a fluff of bits and bytes? Don't count on it."[11] As discussed throughout this book, your likelihood of success will improve dramatically when you appreciate both the strengths—and the limitations—of imaging technology.

---

11.   Kalthoff, "Large Scale Document Automation," 453.

Demonstration Questions and Answers

During the demo, questions concerning the technology will typically arise. Vendor staff should be articulate when addressing the questions asked most often. For example:

Can a document be changed after it is scanned?

If it is stored on WORM (write-once, read many), it can be destroyed but not changed. Erasable and rewritable optical storage is now available, however, that would allow an image to be retrieved, modified and returned to storage. WORM is preferable for archival storage, where image integrity is essential.

How can I tell if a document has been changed?

When expertly performed, it is impossible to determine if an imaged document has been electronically modified. For example, a photograph can be imaged, altered electronically and reproduced without leaving any traces of a "cut and paste."

Where is the document stored?

Permanent storage may be *near-line* at the data server, or *far-line* in a records library. When an image is in active use, however, it is usually stored at the workstation using internal memory, also known as *random access memory*, or RAM.

How many times is a document stored?

It is stored permanently only once, on a centralized optical disk, or jukebox.

Why does it take longer to retrieve some documents?

To optimize performance, frequently used documents may be stored *near-line* while less frequently used documents are stored *far-line*. An image in near-line storage may take 5-60 seconds to retrieve as its optical platter is located and mounted. An image stored far-line may take minutes or hours to access, since a system operator must locate the optical disk manually in a records library.

Can the same document be retrieved by several users simultaneously?

Yes. An electronic copy of the image can theoretically be transmitted to any number of users, virtually simultaneously. The original will remain unchanged on optical storage.

How many ways can one document be indexed?

The only limitations exist in the imaging software's database. With a relational database, the number of indexes may be unlimited for practical purposes. Retrieval time will usually be fastest with the primary (or most frequently used) index, since records may be physically grouped on a disk, based on this index. Images grouped together on a disk can be retrieved with only one "read search," speeding your access time.

Can I search for a document based on any word?

Yes. *Full-text retrieval* systems allow a document image to be searched based on any word or phrase, or combination of words or phrases.

2

The Organization and Imaging

"The most profound technologies are those that disappear. They weave themselves into the fabric of everyday life until they are indistinguishable from it."
— Mark Weiser, Xerox Corporation[1]

Imaging is sometimes called a strategic technology. By contrast, facsimile transmission (fax), although technically similar to imaging in some aspects, cannot be considered strategic. While fax has increased office productivity, it is a simple, inexpensive and utilitarian technology. As with the telephone itself 80 years ago, the adoption of fax by business has been rapid and nearly universal. A business may be hurt by not investing in fax, but no

---

1. Mark Weiser, "The Computer for the 21st Century," *Scientific American*, September 1991: 94.

organization can strongly differentiate itself through this simple technology.

Document imaging systems, however, are neither simple nor utilitarian. Unlike fax technology, imaging does not provide off-the-shelf solutions, but rather an opportunity for unique solutions. Document imaging is analogous to robotics in a manufacturing operation; it represents a new process, rather than simply a new tool. The best results are achieved when workflows are redesigned to exploit this new technology.

As noted in Chapter 1 in the Wang study, organizational barriers represent one of the three major obstacles to effective document imaging. Although technical expertise is needed to design and install these systems, an understanding of business goals is equally important to success.

Strategic Goals

In his book, *Competitive Strategy*, Michael Porter identifies three general strategies for organizational success:[2]

- ♦ differentiation of products or service
- ♦ cost leadership
- ♦ market focus

A business has successfully differentiated its products or services if they are widely perceived to be unique or superior in quality. Cost leadership is achieved when an organization has low manufacturing, distribution or service expense relative to its competitors. Porter terms differentiation and cost leadership *industry-wide strategies*. Following Porter's third strategy, a busi-

---

2. Michael F. Porter, *Competitive Strategy: Techniques for Analyzing Industries and Competitors* (New York: Free Press, 1980).

ness may choose to be more focused, offering a product or service that one segment of the market considers a superior value.

Some businesses pursue a product differentiation and cost leadership strategy simultaneously. Japanese auto manufacturers, for example, do not typically sell the highest or lowest priced cars, but rather offer strong perceived value within the moderate-to-high price range. They have successfully pursued both low cost and quality.

More often, it is a dangerous strategy to be "in the middle." In the retail marketplace in the 1980s, for example, Sears lost significant sales both to low-cost leaders (such as Wal-Mart) and to high-end or niche merchandisers (such as Nordstroms and The Limited).

Support Strategy through Imaging

The most successful document imaging projects begin with a confirmation of corporate objectives and a determination that imaging can uniquely support these objectives. The strength of imaging as a strategic technology is, in fact, its flexibility in support of diverse strategies. While this book is not a primer on corporate strategy, senior management should define a successful imaging project in terms of corporate goals. For example, senior management may state that, "An imaging project will be considered successful if . . .":

". . . customer service is determined to be superior to our major competitors, in our annual quality surveys."

". . . our product's 'design-to-market' time cycle can be reduced by 25 percent."

". . . we can respond with 99.5-percent accuracy to regulatory challenges concerning the documented quality of our pharmaceutical products."

". . . we can locate, by key word, 90 percent of all legal documents considered relevant in our litigation cases, and retrieve these documents within 15 minutes of an attorney's request."

Specific performance thresholds may be drawn from an industry's experience with imaging, and can likely be refined further as a business case is developed (as discussed in Chapter 4). The essential issue is to define the support to be provided by imaging. American Express, for example, traditionally differentiated its credit-card service by returning the signed credit voucher to a customer. Through imaging, American Express successfully reduced the staffing size of its processing department from 300 to six while also controlling postage expense. Imaging was used to maintain product differentiation, in the face of rising costs.[3]

Vendors or consultants can familiarize senior management with effective applications of imaging in their industry. (Figure 2.1 presents typical imaging applications, by industry.) Many industry conferences now feature speakers presenting imaging "success stories." Because imaging systems require a long-term commitment of corporate resources, the importance of senior-management support for imaging cannot be overstated. As with the investment in mainframe systems in the 1960s, or the investment in PCs and LANs in the 1980s, imaging is "infrastructure"—once installed, it requires continual support, maintenance and development. And once integrated with mainframe applications, it may be difficult or impossible to "deinstall."

---

3. "Imaging White Paper: Results from the Imaging Working Group," *Optical Information Systems*, May-June 1990: 144.

Figure 2.1 Potential Imaging Applications, by Industry

Industry	Potential Applications
Banking/Financial	Check processing Signature verification Loan processing and servicing Credit card servicing Mutual funds servicing Trust file management Retirement accounts management
Engineering/Manufacturing	Design (multi-media) Blueprints management CAD/CAM integration
Government	Birth certificates Death certificates Marriage certificates Land records Tax records Court records
Health Care/Pharmaceutical	Patient histories Visual records (x-rays) Insurance records FDA records maintenance
Insurance	Claims processings and servicing Contract management
Law	Litigation support Contract review
General	Payment processing Accounts receivable Human resource information Computer reports (COLD) Archival storage

Select a Leader

At the inception of an imaging project, the primary resource commitment will be time. A task force leader may be devoted full-time or part-time to the effort, based on the scope of the project. Whether drawn from the user group or the information systems (IS) department, the leader should have the perspective of a *sociotechnologist*.

The term sociotechnologist has been used by various integrators and vendors to describe a person equally attuned to technical and organizational issues. A task force leader lacking technical training may be unaware of emerging capabilities in areas such as optical character recognition (OCR) or multimedia integration. Alternately, a task force leader unskilled in issues of job design, team building and employee motivation may advocate technically elegant solutions that employees consider awkward or ineffective.

Document imaging represents an historic convergence of the paperflows traditionally managed by users and the application systems managed by data processing staff. The task force leader is more than a "liaison" between end-users and the systems department. A liaison serves as a translator. An imaging task force leader must serve as a teacher, to help establish a common language of communication. The task force leader should always be an employee of the organization, to ensure project ownership. However, an organization may find it useful to appoint an imaging consultant to the task force, an outsider with the third-party objectivity to foster communication between departments.

Select a Candidate Area

The task force leader and senior management will select one or more areas for further analysis, from the list of candidate areas previously developed. It should not be a foregone conclusion that document imaging is the only technology that may improve these procedures. The relevance and nature of any imaging solution must be established through a detailed evaluation process. For smaller organizations, the first area of potential opportunity for imaging may be easily identified. Larger firms, however, should consider organizational and technical issues when selecting a candidate area.

Organizational Issues

The primary organizational issue when selecting a candidate area for imaging is balancing the risk/return tradeoff. "The most important strategic applications," according to Thornton May, "occur in the customer-service area."[4] These opportunities include processing orders, answering inquiries and handling complaints. The timeliness and accuracy of customer service is usually improved significantly with imaging. Customer satisfaction increased from 68 to 90 on a standardized survey in one organization, following an imaging installation.[5] (Figure 2.2 presents criteria to use in determining high-payback opportunities for imaging.)

The same characteristics that provide a high payback, however, also make them high risk. A failed or awkward installation in customer service affects the lifeblood of your business—your customers. You should weigh strategic benefits against more practical concerns, such as the scope and complexity of the project. A balance of the risk/return tradeoff is advocated by Mary Ann O'Connor. "Each organization has their own goals and objectives to consider," she states, "but I would certainly

---

4.   Thornton May, "Image Processing: Justifying the Image," *Datamation*, April 15, 1990: 82-84.
5.   Ibid.

Figure 2.2 Selection Criteria for Candidate Areas

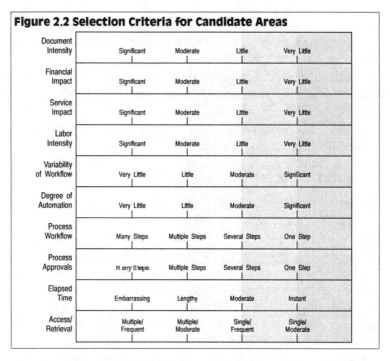

Document Intensity	Significant	Moderate	Little	Very Little
Financial Impact	Significant	Moderate	Little	Very Little
Service Impact	Significant	Moderate	Little	Very Little
Labor Intensity	Significant	Moderate	Little	Very Little
Variability of Workflow	Very Little	Little	Moderate	Significant
Degree of Automation	Very Little	Little	Moderate	Significant
Process Workflow	Many Steps	Multiple Steps	Several Steps	One Step
Process Approvals	M any Steps	Multiple Steps	Several Steps	One Step
Elapsed Time	Embarrassing	Lengthy	Moderate	Instant
Access/ Retrieval	Multiple/ Frequent	Multiple/ Moderate	Single/ Frequent	Single/ Moderate

recommend starting with the least complex problem and going forward from there."[6]

Technical Issues

Technical considerations also influence the risk of a project and perhaps its feasibility. An organization may choose to avoid an extensive multimedia project until stronger industry standards emerge. If your hardware vendor of choice has delayed release of an enterprise-wide imaging solution, you might begin with an electronic file cabinet for archived records that will not need

---

6.   Mary Ann O'Connor, "Applications of Optical Image Technology: Information Retrieval," *Document Image Automation*, May-June, 1991: 166.

integration. If your data network is near capacity, you can begin with a departmental project and pursue an enterprise-wide solution following a communications upgrade.

The maturity of your local area network (LAN) should also be considered. Imaging places significant demands on a LAN environment because of the vast amounts of data scanned and transmitted. No organization should consider a mission-critical imaging application until it has a "bullet-proof" LAN infrastructure—skilled and plentiful LAN staff equipped with proper diagnostic tools. LANs represent, quite literally, the weakest link in the chain for most imaging installations. A multimillion-dollar imaging production system can be brought to a standstill when a 20-cent fuse is blown in the LAN's physical ring. Without skilled LAN administrators and diagnostic tools to quickly isolate and resolve the problem, the battle may be lost for lack of the proverbial nail for the horse's shoe.

The time and effort needed to scan existing documents (called *backfile conversion*) also can be a major operational concern. Departments with a short "document-processing cycle" may benefit from imaging without a backfile conversion—only new documents would be scanned. These departments may only rarely need to access documents more than a few months old. In some departments, however, you may need to convert many years' worth of documents before benefit is achieved. If two projects are both technically feasible and offer similar benefits, the cost of backfile conversion may be a decisive factor.

Select Task Force Members

After a process or department has been nominated for detailed analysis, task force members can be selected. Studies of group dynamics suggest that the most effective groups contain no more than seven or eight individuals. Larger groups limit discussion and dilute responsibility. Task force representatives typically include first-level managers from the department under consideration and information systems (IS) professionals. In larger departments, two or more first-level managers may be needed to provide grass-roots familiarity with detailed procedures. Appointed IS staff should include the proposed systems project leader for any later development effort. The systems project leader will serve as the liaison between the IS department and the task force. In some cases, the task force leader and the systems project manager will be the same individual.

The task force may also include representatives from other departments, such as marketing, internal audit or finance. A task force member from finance, or an internal operations consultant, may be responsible for developing the "business case" (discussed in Chapter 4). A representative from internal audit may offer valuable insights concerning issues of data security and control. And a marketing representative may provide the customer's perspective to the task force. These factors should be considered when selecting task force members:

Creativity

Because the members will be looking for new and better ways to design work tasks, they should be able to step outside of habit and tradition and seek "breakthrough solutions." As described by Stacy Graf, a consultant with Arthur D. Little, "Their goal is to challenge the status quo and derive high-quality solutions, rather than getting 'boxed' in by the existing structure and systems."[7]

Communication Skills

Members of the task force should be well-respected in their individual areas and able to gather support for new initiatives.

---

7.   Stacy L. Graf, "Business Process Redesign: Creating the High-
        Performance Business," Connect, May 1989: 2-5.

They must be able to function effectively within their departments and participate fully in task force discussions.

Internal Status

Although the task force may include personnel from a variety of internal job ranking levels, its members are peers within the task force. Ideally, the group will make decisions through consensus to ensure its final recommendations are "owned" and supported by the entire team.

Accountability

Ideally, the task force leader reports directly to the organization's president, director or CEO, to ensure senior management sponsorship of its recommendations. In larger organizations, the task force leader may report to a steering committee composed of senior executives appointed to oversee a large-scale roll-out of document imaging technology. Of course, there can only be one leader designated for the effort. Whether from IS or the end-user community, the leader must be non-partisan and fully represent corporate interests for the duration of the project.

The Role of the Task Force

For most organizations, quality, service and innovation have replaced cost control and growth as "critical success factors" in the 1990s. Underlying procedures in many organizations, however, still reflect the priorities and assumptions of the 1950s and 1960s. Many tasks remain that provide control at the expense of responsiveness. Merely accelerating these outdated tasks through imaging is the automated equivalent of "paving the cowpaths."

It will be the role of the task force to question basic assumptions regarding a process and its relevance to organizational goals. Task force members will spearhead an effort to "disassemble" current manual procedures, discard irrational or outdated procedures, retain procedures that add value, and "reassemble" these procedures around imaging and other technologies. This effort is commonly called *work design*.

Work Design

The phrase *work design* was coined in the early 1900s by Frederick W. Taylor (Yale professor and father of modern industrial engineering) to describe the scientific review and redesign of manufacturing processes. Work design has been employed throughout this century to structure manufacturing tasks around developing technologies (most recently, robotics, and computer-aided manufacturing, or CAM). Work design has recently enjoyed a resurgence in popularity as *workflow re-engineering*, a phrase most often used when this concept is applied to service industries. A related concept called *information engineering* has also been popularized in books by James A. Martin to describe the redesign of automated systems in support of organizational goals.[8]

It can be stated fairly that the purpose of both workflow re-engineering and information engineering is the *integration of function, process and data*. Function refers to a complete task, such as processing an insurance claim for an auto accident. Process refers to specific activities within the function, such as reviewing a report by an insurance-claims adjustor. Finally, data refers to any information or image needed to support the process, such as the photo of a car.

A flowcharting technique called a Warnier-Orr diagram (after its developers, Michael Warnier and Ken Orr) can be used to depict function, process and data. Warnier-Orr diagrams are

---

8. James A. Martin, *Information Engineering: A Trilogy* (Englewood Cliffs, N.J.: Prentice-Hall, 1989).

similar to the dataflow diagrams or models employed in systems development. However, Warnier-Orr diagrams present a function (that is, both the process and the data) from the end-user's perspective. Dataflow or data-modeling diagrams present data only. Dataflow diagrams, while essential to systems development, consider users only indirectly in the design process. Warnier-Orr diagrams, or similar flowcharting techniques, provide a living, three-dimensional view of the process (Figure 2.3).

Whatever technique is employed, procedures in the department under consideration should be flowcharted. These flowcharts are then reviewed in detail to identify opportunities for improvements in effectiveness or efficiency. The goal is to eliminate unnecessary or redundant tasks, streamline necessary tasks and improve the use of automated systems. In his book, *In Search of Excellence,* Tom Peters calls this approach to productivity improvement "the technology of the obvious."[9] The following principles should be followed when re-engineering these procedures:[10]

♦ Capture information only once, and at its source.

♦ Allow those who capture the data to process it.

♦ Integrate paperflows and data at the beginning of the process rather than at the end.

♦ Delegate decision-making while automating controls.

♦ Provide "do-it-yourself" solutions to enable the staff to resolve problems themselves.

♦ Rethink organizational boundaries.

♦ Organize jobs around corporate objectives.

---

9. Thomas J. Peters, *In Search of Excellence: Lessons From America's Best-Run Companies* (New York: Warner Books, 1982).
10. Graf, *Business Process Redesign,* 2-5.

♦ Consolidate and flatten hierarchies.

♦ Eliminate steps that do not directly add value to the customer.

Workflow Diagrams

Warnier-Orr diagrams are constructed from left to right and from top to bottom. Major functions are listed on the far left; using brackets, specific activities within those functions are listed in chronological order. Activities are often broken down into sub-tasks, again, using the brackets.

The strength of the Warnier-Orr technique is its flexibility, allowing any level of detail to be presented without the need to reconstruct a diagram. When further detail is needed, additional brackets are drawn and the diagram expands from left to right. On the far right, specific data or images required in the process can be identified.

From Work Design to Imaging

As a process is being re-engineered, the capabilities of imaging (and other technologies, if relevant) should be considered. As a prelude to the development of technical requirements needed in a Request for Proposal to vendors, you should address the following questions:

What documents should be stored as images?

The task force must identify the documents to be scanned and stored as images. As noted previously, if electronic data capture (EDC) or electronic data interchange (EDI) are infeasible, externally generated documents are likely candidates for imaging. Depending on the nature of the business, these documents can include customer orders, claims, supplier invoices, resumes, general correspondence or legal

Figure 2.3 Warnier-Orr Diagrams

Check-Cashing Procedures in a Non-Imaging Business Environment

Cash On-Us Check
- Verify Customer Completed Check Correctly
  - Date
  - Payee
  - Amount
  - Signature
  - Endorsement
- Verify Funds Available from Teller Terminal
- Verify Customer Signature
  - Walk To Signature Card File
  - Retrieve Signature Card
  - Verify Signature
  - Refile Signature Card
  - Walk Back To Teller Window
- Validate Check
- Give Customer Cash

Check-Cashing Procedures in an Imaging-Supported Business Environment

Cash On-Us Check
- Verify Customer Completed Check Correctly
  - Date
  - Payee
  - Amount
  - Signature
  - Endorsement
- Verify Funds Available from Teller Terminal
- Verify Customer Signature from Teller Terminal (automatically displayed)
  ValidateCheck
  Give Customer Cash

Source: Edwin D. McDonell, "Corporate Reeningeering that Follows the Design of Document Imaging." *Information Strategy: The Executive's Journal*, Fall 1991: 7.

depositions. The only factors shared by these documents is the arrival of this paperwork from a third-party and an inability to automate the transaction by other means.

After you establish that a document should be converted to image, a related question is how many of the older documents or backfiles should you convert. The cost of backfile conversion can be significant. Many organizations choose to convert documents only as they are needed. Prior medical records, for example, might only be imaged if the patient returns to the physician or the hospital.

Archived documents are also candidates for imaging. Government agencies in the United States and abroad are active users of imaging for the storage of vital records (birth certificates, marriage licenses, death certificates) and other archived materials. Firms in heavily regulated industries—such as pharmaceuticals, insurance and finance—also have significant archival needs. Manufacturers claim a shelf-life of 30 years or more for records stored on optical disk. By contrast, microfiche tends to deteriorate after 7-10 years. It should be remembered, however, that data stored on optical disk is only as long-lived as its supporting technology—the software and hardware needed to read and retrieve the data. Emerging technical standards, discussed in Chapter 5, are therefore an important consideration in the selection of an imaging system.

In general, internal forms and documents should be incorporated into the application system whenever possible, rather than being handled as images. A loan application form, for example, may be recreated as a data-entry screen on the loan-processing system. Loan officers would then enter data directly into the application system, eliminating any paper-based process. It is occasionally desirable, however, to image internal forms. A Texas-based lessor of gas compressors, for instance, chose document imaging for maintenance reports submitted via fax from remote locations. The alternative—mobile, laptop PCs installed in trucks and linked by cellular communication—was considered more problem-prone and expensive.

Finally, system-generated hard-copy reports can be provided through computer-output to laser disk or (COLD). With COLD, system-generated reports are available for viewing on-line. COLD eliminates the need to print, burst and distribute reports while providing faster access than the traditional alternative of microfiche. In a variation of COLD, access to manuals or directories is provided via optical storage on portable CD-ROM disks (virtually identical to music CDs). CD-ROM combines the advantages of fast, indexed retrieval of significant quantities of data with inexpensive distribution costs. New CDs are simply mailed when manual or directory updates occur. A disadvantage of CD-ROM is the high initial production cost, although the unit cost is low if large quantities are manufactured. To access the CD, a PC or server must also be augmented with a CD reader.

Where in the organization will these images be used?

Once it is determined that a document should be imaged, the departments involved in the review or processing of the document should be identified. For work-in-process documents, the image may need to be forwarded between departments as it is reviewed, requiring LAN communication between workstations. Archived images, by contrast, may be provided through one workstation in a single department.

Any decision-making process should be identified (for example, "...if a signature is missing, return the document to Customer Service") so that it can later be incorporated into the automated workflow of the imaging system. Workflow logic distinguishes advanced imaging systems from simpler electronic file cabinets. With workflow software, processing tasks can be prioritized, workloads can be equalized, and completion dates can be estimated. In a sense, the imaged document is programmed to manage its own forwarding, review and disposition. Imaged documents should be scanned as closely as possible to their site of origin. Ideally, the first person handling the document should scan it and, as discussed later, index it. All subsequent access to the document will be via image.

Who should be authorized to store, access or modify the images?

Departments need to consider access to images by specific individuals as well as the type of access provided. The number of authorized parties will directly influence the number of display workstations needed. To avoid an over-investment in technology, estimate the anticipated users based on anticipated productivity gains. As with traditional automation, you can define access procedures to allow:

♦ image entry only

♦ image inquiry only

♦ image modification

♦ combined authority for two or more of the preceding capabilities

The nature of access should be defined on a "need-to-know" basis. Clerks operating scanners only need the capability to view their own documents to ensure proper scanning. Other clerks may only be allowed to index documents. Processors may have inquiry access only to specified documents (or even to just a portion of a document). Managers may have only inquiry capability. Annotation capabilities can be provided to some processors or managers.

This can include screen annotation ("writing on" the document, using an electronic stylus and pad) or voice annotation (describing an issue or concern into a phone, for later play-back).

Finally, the capability to modify the document image itself can be provided under certain circumstances. Image modification most often occurs in special imaging applications, such as when an artist "ages" the scanned photograph of a missing child to convey his or her likely appearance. With WORM (write-once, read many) optical storage, an image can be destroyed but not modified. Now gaining in popularity, rewritable optical storage expands the opportunity for fraud—such as the superimposing of a signature on a document—unless strict controls exist. As discussed in Chapter 6, the admissibility of an image as a legal record depends on your organization's capability to prove that the original image has not been altered.

When will these images be needed?

Frequency of access should be used to determine the *speed of retrieval* required. The cost of document imaging increases exponentially based on the speed of retrieval demanded. The least expensive storage, called *far-line*, is provided by an optical disk stored in a records library. When an image on the disk is requested, an operator will physically mount the platter on an optical drive (in the same way a CD is loaded in a stereo). The retrieval time for far-line storage may be one minute or several hours.

Near-line storage refers to image retrieval from a disk mounted in an optical jukebox. When the image is requested, a mechanical arm in the jukebox physically mounts the platter for reading (in a manner not unlike a 45 rpm record in a music jukebox). Jukeboxes require from 5-60 seconds for image retrieval. With jukeboxes, retrieval speed is fastest when the record to be accessed is on a disk already mounted in a drive. Jukeboxes for 5.25-inch disks have from one to 16 drives and hold a maximum of from 20 to 512 disks. Jukeboxes for 12-inch disks have from four to 112 drives (when daisy-chained together) and hold a maximum of from 64 to 2,400 disks.[11]

On-line storage refers to image retrieval from local storage. For fastest retrieval, the image may be stored in *random access memory* (RAM) or on *magnetic disk* (hard disk or diskette) at the display workstation, or it may be stored on optical disk at the display workstation, or at the system's data server. Images stored on-line can be retrieved in five seconds or less.

How will the images be accessed and used?

Flexibility of access is imaging's great advantage over paper-based solutions. With paper-based files, one indexing method is typically selected (such as account number or customer name). Any additional index requires that all documents be duplicated, literally doubling file maintenance costs. The heart of many image-management systems is a relational database. This database is expanded to include an image and its indexes as data elements. Thus enhanced, the database can be used to retrieve an image through any index established when the document was scanned. A purchase order might therefore be retrieved by customer, account number, date sent, date received or processor.

Efficient but accurate indexing is an important issue in the design of an imaging solution. Optical character recognition (OCR) combined with proper forms design may

11. Bailochan Behera, "Jukebox Cost Performance Modeling," *Document Image Automation*, May-June 1991: 152.

permit a document to be indexed automatically as it is scanned. If handwritten data always appears in the same area on a form, and it can be interpreted by OCR, manual indexing is eliminated. The current database system can also be used to facilitate indexing. In the case of a purchase order, the account number may be entered manually; however, the customer's name and the date processed might then be assigned automatically, assuming this data is linked within the application system to the account number. The application system may also help *validate* the index data assigned (such as confirming that an account number is valid by comparing it with a table of current account numbers in your application system).

Gather and Organize Data

The preceeding questions can only be answered through a detailed analysis of your operations. For the candidate area, your task force should gather and organize data on current procedures. Workflow diagrams can first be created detailing manual activities. These diagrams will then be reviewed to identify opportunities to improve the process through imaging. The following steps are suggested:

♦ Distribute survey questionnaire.

♦ Interview potential end-users.

♦ Create workflow diagrams of the original process.

♦ Create workflow diagrams of the image-enhanced process.

♦ Share the new process with department for comments.

Survey Questionnaire

A questionnaire can be used to gather preliminary information concerning customer-service issues, processing bottlenecks, workload distribution and workflows. If your task force cannot interview all staff in a department, survey questionnaires will help direct your efforts (Figure 2.4).

Interviews

Your task force should interview in detail at least one employee within each job category. If a department has 10 accounts-receivable clerks, for example, you should interview at least one, and preferably two. The interviewer should be "walked through" the person's daily work. If customers are involved, the interviewer may play the role of the customer to simulate the transaction. During the interviews, discussions should include report distribution and use, forms and supplies utilization, automated equipment needs and physical space limitations. Areas where work becomes bottlenecked, or where necessary documents or supplies are not immediately available, should be noted.

In these interviews, you should identify not just what is done, but also why it is done. It may be valuable, for example, to trace the path of Form 74B-83 from department to department, but it is more important to determine why Form 74B-83 exists in the first place. Does it directly add value, or have you enshrined a procedure that has outlived its relevance?

Why does a supply requisition for a box of pencil lead require two signatures—because a president who was trying to cut expense created this procedure 10 years ago? If the president has since been fired by the board of directors and the cost of handling the form exceeds the value of the pencil lead, you have "institutionalized the irrelevant," in the words of management consultant Michael Hammer.[12]

---

12. Michael Hammer, "Re-engineering Work: Don't Automate, Obliterate," *Harvard Business Review*, July-August, 1990: 104-112.

Figure 2.4 Survey Questionnaire

Name _____ Title _____

Department_____ Date _____

1. How many hours per week do you spend on the following tasks?

    a. searching for documents _____
    b. creating files _____
    c. maintaining files _____
    d. making copies _____
    e. distributing documents _____

2. How often has the lack of a document forced you to delay a business activity or decision?

    _____ times/day or _____ times/month or _____ times/year

    Describe example circumstances _____
    _____
    _____

3. Have any important documents you needed ever been completely lost?   Yes _____ No _____

    If yes, describe example circumstances _____
    _____
    _____

4. What documents do you use most often?

    a._____ b. _____ c. _____
    d._____ e. _____ f._____

5. Where do these documents originate? (answers for questions 5 through 12 should correspond to the documents listed in 4).

    a._____ b. _____ c. _____
    d._____ e. _____ f._____

Figure 2.4. Survey Questionnaire (continued)

6. How are these documents organized so they can be located? (i.e., what indexes—such as date, customer name, or invoice number—are used to locate these documents?)

    a. _____ b. _____ c. _____
    d. _____ e. _____ f. _____

7. How would you like these documents organized, if you could have another index or several indexes?

    a. _____ b. _____ c. _____
    d. _____ e. _____ f. _____

8. What activities do you perform with these documents? (select as many of the following as apply: 1) retrieve, 2) review for information, 3) review for decision making, 4) process, 5) route, 6) file.

    a. _____ b. _____ c. _____
    d. _____ e. _____ f. _____

9. How are these documents disposed of when they leave your desk? (choices: 1) filed/personal file, 2) filed/department file, 3) routed, 4) discarded)

    a. _____ b. _____ c. _____
    d. _____ e. _____ f. _____

10. If the document is routed, where is it routed?

    a. _____ b. _____ c. _____
    d. _____ e. _____ f _____.

11. How are routed documents tracked? (select as many of the following as apply: 1) with automation, 2) with file folder, 3) with file log, 4) with automation, 5) other—describe).

    a. _____ b. _____ c. _____
    d. _____ e. _____ f. _____

Figure 2.4. Survey Questionnaire (continued)

12. Is access to any of these documents restricted? (choices: 1) to you personally, 2) to department, 3) to supervisors, 4) to management, 5) other—describe).

a. _____ b. _____ c. _____

d. _____ e. _____ f. _____

Procedural Diagrams

Through the questionnaires and interviews, the task force can construct a diagram of current procedures. Task force members at one Banc One unit, for example, in their review of a large department, literally "papered" the wall of their meeting room with procedural diagrams. They then marked off in red pencil any procedures that were simply redundant or irrelevant. They also marked off in blue pencil all procedures that could be supported through imaging and other automation. This "annotated" diagram was then shared with the department under review (a student loan processing group) for comments and suggestions. Through this input, a new diagram was created to define how the process would work when automated.

Don't Pave the Cow Paths

Once the task force has identified a new process using imaging, you should reconfirm that your solution supports the goals defined earlier by senior management (such as improved customer service or faster product speed-to-market). The most important question—"Why will document imaging enhance our organization's performance?"—can be revisited as a touchstone throughout the project. According to Michael Hammer, writing in the *Harvard Business Review*, "Companies tend to use technology to mechanize the old ways of doing business. They leave the existing processing intact and use computers simply to speed them up."[13]

To Hammer, applying new technologies to outdated procedures is "paving the cowpaths." He relates the story of Ford Motor Company's efforts to streamline and automate their accounts-payable department. By automating existing manual

---

13.   Ibid.

tasks, they estimated that their headcount in a 500-person department could be reduced by 20 percent. They were enthusiastic about their plan—until they learned that Mazda's accounts-payable department consisted of only five people. Ford recognized through further scrutiny that its accounts-payable process was basically flawed—they paid suppliers when *invoices* were received. The company fundamentally revised the process to issue payment when *goods* were received—thus eliminating a vast effort to reconcile the paperwork associated with invoices and shipping documents.

By questioning basic assumptions, Ford achieved what Hammer calls a "breakthrough solution." As Hammer writes, "We cannot achieve breakthroughs in performance by cutting fat or automating existing processes. Rather, we must challenge old assumptions and shed old rules that have made business underperform in the first place." If your organization is willing to question and challenge tradition, imaging can provide a new way of doing business.

3

Communicating with Vendors

"Vendors estimate that more than 40 percent of RFPs (Requests for Proposals) are so poor in quality they cannot be responded to effectively."
    –Scott McReady, International Data Corporation[1]

Imaging systems bring a special challenge to the procurement process. Most imaging solutions are complexes of hardware and software uniquely developed for each installation. They likely require one-of-a-kind application software, typically developed by the vendor rather than by your IS department. In addition, software and hardware may be from several sources. The list of potential vendors is diverse, including systems integrators, application-software vendors and manufacturers of specialized hardware, such as jukeboxes.

---

1.   Interview of Scott McReady by Kathleen Hawk, 1992.

You should consider an imaging vendor a potential consulting partner rather than a product supplier. When selecting a vendor, you may contract for services—including project management and training—as crucial to your success as the technology itself. Even after the contract is signed, the project may be measured in years, not months. And it can be assumed that the scope of the system will expand as you discover additional opportunities to employ the technology. These circumstances dictate that the purchase agreement be as clear as possible. Expectations on both sides should be precisely detailed. The more knowledge both buyer and seller have concerning the nature of the business problem and the capacity (and limitations) of the technology, the greater the likelihood of success.

The RFP Defined

The Request for Proposal (RFP) is a formal bidding document that provides a framework for the diverse issues addressed in the procurement process. The RFP organizes communication between vendors and buyers through detailed functional and technical specifications. Although RFP development is a time-consuming effort, it prevents a "ready, fire, aim" approach to system expenditures. Benefits of a formal Request for Proposal include the following:

- *Accountability by end-users in determining business requirements.* No vendor understands crucial organizational and competitive issues as fully as you, the buyer. Buyers that take a passive role and accept unsolicited vendor proposals are avoiding their own responsibility to understand and define business needs. The involvement of your end-users also establishes accountability and ownership, and ownership is essential to the success of the project.

♦ *A competitive bidding process between vendors.* The RFP is a detailed definition of needs, allowing vendors with diverse solutions to bid appropriately and be compared against each other on an "apples-to-apples" basis. It is a formal, objective framework that provides a level playing field for vendors while ensuring a cost-effective solution for the buyer.

♦ *An opportunity for creative insights from multiple vendors.* Each vendor or integrator will have a unique perspective concerning potential solutions to your business problem. The strongest vendors will seek to offer creative and cost-effective solutions. Their insights may add to the value of the system designed and installed, regardless of which vendor you select.

♦ *A document sufficiently precise to be contractually binding.* A vendor's response to the Request for Proposal can later be included as an addendum to your contract. These detailed representations concerning system functionality and support can be linked to performance guarantees and payment holdbacks, to provide you with leverage after the contract is signed.

These benefits more than compensate for the time required to prepare the RFP. Although a formal RFP may add two or three months to a project, your only real options are "do it right" or "do it over." More than one organization has had to deinstall an imaging solution that could not address its needs, or renegotiate a contract that was hastily signed. Properly developed, the RFP will help ensure that you "do it right" the first time.

Vendor Involvement

System specifications cannot be defined without a clear understanding of the buyer's needs. According to industry analyst

Scott McReady, vendors find many RFPs unanswerable because they are vague and poorly defined, or are unrealistic in their expectations.[2] Your detailed functional requirements are the foundation of the RFP: types of documents to be captured and stored, number of users, retrieval requirements and document-retention schedules (see –Request for Information on page 84).

Vendors then respond with automated solutions matching your needs. As you define your needs, a vendor's perspective may be useful. Vendor involvement likely began with the education of end-users during imaging demonstrations. It may have continued if an integrator assisted in a work re-engineering study. It may be useful to continue this participation by sharing a Request for Information (RFI) with one or more vendors before you issue your Request for Proposal. A Request for Information is a more limited document that can help you determine whether your needs can be met through imaging, and can help identify where additional detail may be needed in your RFP (see Request for Information on page 86).

Structure of the RFP

A Request for Proposal is organized into logical sections to facilitate review and response. Although there are many valid formats for an RFP, an AIIM technical committee has developed guidelines and an example RFP that can be adopted and modified as desired.[3] Our suggested RFP structure expands on this AIIM format and includes the following sections:

♦ Introduction

♦ Proposal Administration Section

---

2. Ibid.
3. Association for Information and Image Management, *Electronic Imaging Request for Proposal (RFP) Guidelines* (Silver Spring, Md.: AIIM, 1991).

- Requirements Section
- Pricing Section
- Contract Section

Introduction

The introduction presents background information to familiarize a vendor with your business. It typically includes the following:

- an overview of your organization
- your organization's business strategy
- your information systems environment

Overview of the Organization For public corporations, this description can be drawn from your annual report and will include data such as services provided or products manufactured, annual revenues, asset size, number of employees and office locations. For government entities and not-for-profits, any authorized "statement of purpose" can be offered along with data relevant to your organization's mission—a hospital, for example, may cite its number of beds and annual outpatient volumes. If a departmental solution is sought, additional detail should be provided. Organizational charts can help clarify roles and responsibilities.

Business Strategy A statement concerning your organization's competitive strategy should be offered. The planned role of imaging in supporting performance can also be stated. As previously noted, specifics may be developed concerning performance objectives, such as the following:

- Reduce by 30 percent the time needed to respond to customer inquiries.

♦ Achieve a retrieval accuracy of 99.5 percent for archived records.

♦ Approve and process consumer loans within 24 hours.

Current Information Systems Environment An overview should be provided of your technology environment and strategy. Required compliance with specific vendor architectures (such as IBM's System Application Architecture) should be noted. Avoid imposing unnecessary technical constraints on the vendors; however, identify any mandatory requirements. General strategies should be described, such as *right-sizing* your computing environment with client/server technology. Any automation projects underway that will influence the timing or scope of an imaging effort should also be identified.

Proposal Administration Section

This section describes the administration of the proposal process. It provides "ground rules" for the vendors, that include the following:

♦ name, address and phone number of the vendor liaison

♦ proposal due dates

♦ date and location of bidder's conference (if scheduled)

♦ joint bids

♦ required proposal format

♦ vendor evaluation criteria

♦ planned decision date

♦ contractual obligation

♦ confidentiality

Vendor Liaison You should insist that your appointed contact (often the task force leader) be the only channel of communication. A vendor's sales staff is trained to "sell high and sell
wide"—that is, establish a network of contacts with senior management and throughout the buyer's organization. Insiders communicating with a vendor may be inclined to "second guess" the
efforts and decisions of the task force. If a vendor believes it is
losing ground, its representatives may encourage these internal
politics in an attempt to derail the study.

The primary value of a single contact, however, is to ensure
fairness to all vendors. It offers consistency of information to and
from your organization. Outgoing information includes changes
in the RFP that may develop or clarify issues in response to
vendor questions. All vendors, in fairness, should receive the
same information at the same time. This is only accomplished
when one person serves as liaison. This requirement should also
be communicated in your organization, to preempt any vendor's
efforts to do an "end run" around an appointed contact.

Proposal Due Date Vendors typically need four to six weeks
to prepare a proposal. If you contact vendors in advance, however, they can anticipate the RFP and schedule their resources
accordingly. A vendor may take several days to determine who
its representative should be. Experienced buyers will call in
advance, establish a rapport with the marketing representative,
and discuss any concerns the vendor may have with the proposal
due date. If a due date needs to be delayed to accommodate one
vendor, protocol requires that all vendors be granted an extension.

Bidder's Conference Although not mandatory, a bidder's
conference can improve your communication with vendors by
resolving any initial questions or concerns. This conference usually is scheduled four or five days after an RFP is issued. The task
force provides an overview of the project and answers any

questions vendors have concerning the project or the RFP. Vendors with more RFPs than they can answer have an opportunity to meet you face-to-face and determine if a response is warranted. Following the conference, copies of vendor questions and your answers should be sent to all vendors for the benefit of those not in attendance.

Joint Bids Any limitations you want to place on joint ventures should be defined. A vendor may seek to collaborate with an integrator or consulting firm in performing a project. This should be acceptable to you, assuming that a single contractual party with ultimate responsibility is identified. The lead vendor or contractor will then subcontract portions of the work to other providers, while maintaining full responsibility for work quality and delivery dates. Your prime contractor may be an integrator, or may be a vendor providing the majority of the hardware and software.

Required Proposal Format All vendors have off-the-shelf proposals they modify and issue unless a specific format is requested. It is virtually impossible to compare proposals submitted in diverse formats. To permit page-by-page proposal comparisons, you should insist that the vendors follow your requested format. You may want to provide vendors with an electronic template of the proposal format (that is, in ASCII code). They then can convert the ASCII text to their own word-processing software and "fill in the blanks."

Vendor Evaluation Criteria To ensure the best proposals possible, share your evaluation criteria with the vendors. As discussed in the following pages, evaluation criteria can be classified as functional or management-related. Every buyer has a unique perspective when balancing these concerns. Explicit, quantitative criteria for evaluating vendor alternatives can make

Figure 3.1 Vendor Evaluation Criteria

Issue	Technical Criteria	Potential Points
What documents should be stored as digitized images?	Image Quality/Resolution	5,000
	Storage Media	5,000
Where in the organization will these images be used?	Connectivity/Interoperability	5,000
	Communications Capability	5,000
Who should be authorized to store, retrieve or modify the images?	Security/Recovery Features	2,500
When will these images be needed?	Capture/Retrieval Speed	5,000
How will they be accessed and used?	Software Features/Functions	10,000
	Ease of Use	10,000
	Customization Capability	7,500
	Integration With Existing Equipment/Software	10,000
	Expansion Capabilities Without a Conversion	5,000
	Total - Technical Criteria	70,000

Issue	Management Criteria	
What is the vendor's experience and financial stability?	Company History/Longevity	2,000
	Financial Performance	4,000
	Installed Client Base	5,000
	Source of Capital	3,000
What support services are provided by the vendor?	Project Management	3,000
	Conversion	2,000
	Training	4,000
	Customization	4,000
	Documentation	3,000
	Total - Technical Critera	70,000
	Total - Management Criteria	30,000
	Total - All Criteria	100,000

this process as rational as possible. Figure 3.1 reflects one organization's priorities. Of 100,000 available points, 70,000 points have been assigned to functional criteria and 30,000 to management criteria. A vendor that fully met functional requirements would be awarded 70,000 points. Evaluation methods are discussed further in Chapter 5.

Planned Decision Date In fairness to vendors, a planned decision date should be cited and adhered to. A vendor's time is a free commodity to you as a buyer, but not to the seller. Vendors respect organizations that have the capability to make a timely decision and the resolve to act. And when a vendor's proposal is rejected, you should provide the vendor's representative with an explanation so that they can learn from their efforts and improve their chances with future prospects.

Contractual Obligations Advise the vendors that responses to the RFP will be referenced in the contract and will be considered contractually binding. This caveat restrains a vendor's natural tendency to oversell its capabilities.

Confidentiality An RFP often includes sensitive information concerning an organization's current performance and competitive strategy. Include a statement requesting that all RFP materials be used only for the purpose of preparing a proposal. Some organizations attach a non-disclosure statement to the RFP for the vendor to sign and return.

Requirements Section

The requirements section is the heart of the RFP. It contains all the information necessary for a vendor to prepare a proposal. Requirements are typically grouped into the following categories:

◆ *Functional requirements,* describing the capabilities to be provided by the document-imaging system and any technical specification or constraints relating to the hardware or software.

♦ *Management requirements*, in areas such as vendor stability and vendor support.

Functional requirements are the objective criteria in the vendor-evaluation process. Vendor responses to these needs can be readily verified—the vendor can or cannot provide a capability. *Management requirements*, although equally important, are subjective. They include intangibles such as the buyer's confidence in a vendor, or the buyer's opinion concerning software documentation.

Functional Requirements Functional requirements were established in the needs-analysis process, as discussed in Chapter 2. These requirements were established by your response to several basic questions concerning imaging. In the RFP, your responses are summarized for the vendors:

Why will document imaging enhance the performance of our organization?

This question links technology to strategy. Is the system intended primarily to improve productivity or to enhance customer service? Business objectives should be presented in a paragraph or two. This also can serve as an introduction to any internal *business case* needed for imaging (as discussed in Chapter 4). A flowchart of current procedures, and a proposed flowchart of imaging-enhanced procedures, should serve as exhibits.

What documents should be stored as digitized images?

As shown in Figure 3.2, documents to be imaged should be identified. Physical considerations include the sizes of

Figure 3.2 Documentation Information Required When Preparing RFP

Physical Documentation Characteristics

- Percentage of documents by size (in inches)
  - _____ 3×5
  - _____ $5\frac{1}{4} \times 7$
  - _____ $8\frac{1}{2} \times 11$
  - _____ $8\frac{1}{2} \times 11$
  - _____ 11×17
  - _____ Other
- Percentage of documents two-sided
- Type of paper
  - _____ Standard bond
  - _____ Standard paper weights
  - _____ Glossy
  - _____ White
  - _____ Non-white — what colors are used
  - _____ Shading present — percentage
- Type of printing on document
  - _____ Preprinted form
  - _____ Handprinting/handwriting
  - _____ Ink color specified for hand notations
- Physical input of forms
  - _____ Original
  - _____ Photocopy
  - _____ Carbon, which copy is used (2nd, 3rd, etc.)
  - _____ Carbonless press copy
  - _____ Other

Physical Document Input Volumes

- Total number of documents received per day?
- Separate documents by types and numbers
- Document receipt times?
- Are there peak times?
- Are there peak calendar days?
- Number of folders created per day?

Physical Document Storage

- When are folders archived and not readily accessible?
- How often are archived folders accessed?
- Are folders/documents microfilmed?
- What type of film? How is it stored?
- What is the per day number of documents filmed?
- Are documents one- or two-sided?
- How is the film indexed?
- What is the access time?
- How many per day are printed?
- What are the legal requirements for retention?
- When are documents removed from active files?
- How are inactive documents stored and indexed?
- When are inactive documents stored?
- Is there a process for purging inactive documents?

Conversion Data

- Type of material to be converted
  - _____ Paper
    - _____ Sizes of paper
    - _____ Types of paper
  - _____ Microform
    - _____ Roll film
    - _____ Microfiche
    - _____ Aperture cards
- Quantity of material to be converted by type and size
- Indexing methodology for archived material
- Condition of material by type and age
- Current access rate to archived material
- Average time material is out of files
- Internal facilities available
- Personnel available
- Unique problems

Source: Scott Wallace, *Implementing Electronic Imaging: A Management Perspective.* Arlington, MA: Cutter Information Corp.: 1991.

documents, whether they are one-sided or two-sided and whether they are bound or unbound. Volume considerations include the frequency of new documents (stated daily, monthly or weekly), any desired backfile conversion of older documents, and document-retention requirements. This information is essential to a vendor in determining the input and storage requirements for a proposed solution.

It is strongly advised to include a complete selection of the forms to be imaged. Not all forms are "scannable." Although legible to the human eye, black print on a gray shaded background, for example, may be unreadable by a scanner. The cost of forms redesign and redeployment may be significant, and in some instances it may be infeasible to redesign a form (such as when its design is regulated, or when it follows an industry standard). Also, if a form lacks sufficient weight (that is, if it is flimsy), it may affect your ability to employ a sheetfeed scanner. Vendors should be asked to identify any concerns through testing, as they are preparing your proposal.

Where in the organization will these images be used?

The location of end-users must be identified so that the vendor can propose an appropriate software and communications network solution. A departmental imaging solution may be supported by one local area network (LAN), while an enterprise-wide solution will dictate several LANs, or perhaps a wide area network (WAN) that is geographically dispersed. Remote workstations can be supported via dial-up or a dedicated telecommunications line, depending on planned usage. For occasional users in branch offices, support via modem over existing telecommunications lines may be feasible. When remote users do not require real-time data and images, a less expensive option may be to express mail to them duplicate optical platters that can be accessed locally.

Who should be authorized to store, retrieve or modify these images?

The anticipated number of document processors and end-users must be identified, so that vendors can determine workstation and communications needs. The number of processors and end-users, as well as their required work-stations, also dictate the required capacity of the server (that is, its processing speed and storage requirements) in a client/server environment. End-users can be considered the *consumers*, while processors are the *providers*. Processors scan documents, view the image to ensure that it was captured properly, and index the image. End-users may be managers who only view documents occasionally, or clerical staff who employ the images constantly in their work. For both end-users and processors, the organization should define the number of staff required, the number of image retrievals anticipated (on an hourly or daily basis), and the documents (with number of pages) to be retrieved. Vendor assistance may be useful in determining the appropriate number of processors and their support requirements.

When will these images be needed?

Speed-of-retrieval needs must be defined for the vendor. Depending on the business situation, access needs may vary from a second or two to several days. For better or worse, PCs have accustomed many users to immediate data access. One imaging buyer, for example, insisted that only a response time of three seconds or less was acceptable. At their branch offices, this need for three-second access would have increased their system costs by more than 25 percent when compared to a solution offering 10- or 15-minute access. With their current manual system, however, document retrieval for their branches took 24 hours. Because the imaging investment could not be cost-justified with a response time of three seconds or less, the buyer nearly created an all-or-nothing situation in which they were the losers.

Consider the time value of this information. What added value is provided if the document can be retrieved in two seconds rather than two minutes or two days? As discussed in Chapter 2, image storage may be on-line, near-line or far-line. On-line provides the instant access at the greatest cost. Near-line and far-line are slower but less costly options (Figure 3.3). You should seek a retrieval speed that offers the best return on your investment, of course, based on the *time value* of the image. This concept is discussed further in Chapter 4. An initial performance goal must be stated in the RFP, but an optimal retrieval speed may only be determined through a cost-benefit comparison that includes vendor pricing. You may therefore want to request multiple cost quotes, reflecting varying retrieval speeds.

How will the images be accessed and used?

Your planned index or indexes for a document should be defined for the vendors. In some archival applications, only one index is required, such as customer name. For a few applications (such as litigation support) you may need image retrieval using any word or phrase in the document. When imaging is integrated with the application system, cross-indexing may be provided automatically—if a social security number is the primary index, for example, the application database might provide a secondary index by customer name or address.

Figure 3.3 Document Access Options

	On-line	Near-line	Far-line	Off-line
Storage medium	magnetic disk	volume in jukebox	not in jukebox	paper or film
Image is indexed	yes	yes	yes	no
Image on screen	yes	yes	yes	no
Speed of retrieval	less than 5 seconds	5 to 60 seconds	1 to 30 minutes	days
Human Intervention	no	no	yes	yes

Source: David Black. "Is On-Line Good Enough?" *Inform*, January 1990.: 12.

A vendor may offer the capability to establish a new index later, if the need emerges. Imaging systems sometimes employ a *runtime* version of well-known database software, such as Informix or Oracle, to manage the indexing function. In that situation, the inherent strengths and weaknesses of a database can be assessed by IS staff familiar with these solutions.

You also should consider the feasibility of indexing with optical character recognition (OCR). With OCR, scanned documents are indexed automatically, based on characters or symbols in the document. Inexpensive software exists that will recognize symbols (such as bar codes) or characters from any standard font. More sophisticated software, using an emerging technology called *neural programming*, can read handwritten data and characters from unknown fonts.

A question related to indexing is planned usage. What documents will be retrieved and processed together? A vendor should know, for example, that all documents related to a customer's insurance claim are likely to be retrieved and reviewed at once. Retrieval time for these documents can be reduced by 50 percent or more if they are stored in the same *sector* of the optical disk—similar to physically filing all claim documents by customer, so only one file search is needed.

You also should identify special processing requirements. These typically include the capability to annotate an image, route an image automatically, or integrate the image with an existing application. The issue of integration with existing data processing systems is especially crucial. "If integration concerns are not addressed explicitly and early in the requirements definition for the system," states imaging consulting Theresa Elms, "you're liable to be left with a stand-alone application that will never be integrated."[4] As discussed in Chapter 2, procedural diagrams, dataflow diagrams and data modeling can help clarify your requirements.

What are the technical issues or concerns?

Technical concerns may be related to *system performance* or *system architecture*. Performance requirements include issues such as scanner speed or display resolution. System architecture includes issues such as support of specific operating systems or communication protocols. In general, novice buyers should minimize performance-related technical specifications. If your functional needs are fully defined, your performance requirements will follow.

---

4.   Theresa Elms, "Image Processing: A New Vision for Information Systems," *ComputerWorld*, December 12, 1989: 28.

Predefining performance criteria may deprive you of a vendor's expertise. An open-ended strategy enables you to weigh the approach of each vendor and perhaps negotiate a solution that is a hybrid of each vendors' best insights. For example, if you dictate a need for one high-speed 17 page-per-minute laser printer, a vendor may not propose a solution that could be better—three 8 page-per-minute printers. Three slower laser printers would handle the same throughput, while providing greater redundancy in the event of failure.

System architecture includes mandatory issues of technical compliance. Significant effort can be saved by eliminating vendor responses that do not comply with mandated technical strategies concerning operating systems, application program interfaces, communication protocols or software development tools. At a meeting with one buyer, imaging analyst Bud Porter-Roth observed there was a virtually unanimous commitment to OS/2 and 5.25-inch optical media, indicating this group's strong relationship with IBM.[5] If you have a corporate commitment to one hardware provider or technical environment (for example, UNIX), let the vendors know. However, the more detailed an RFP is concerning choices of hardware and software, the less latitude a vendor has in responding to your needs. The general wisdom is to provide as few constraints as possible, while remaining within the scope of your organization's IS strategy.

System architecture also includes issues of system expandability, modularity, ease of maintenance, and ease of integration with existing applications. These topics usually are addressed in other RFPs generated by your IS depart-

---

5.   Interview of Bud Porter-Roth by Kathleen Hawk, 1992.

ment. More than likely, IS staff on your task force can borrow much technical criteria from existing IS documents to employ in the imaging RFP.

Management Requirements Management requirements address qualitative issues of vendor capability, including the following:

Vendor experience and stability

- ◆ company history
- ◆ financial performance
- ◆ personnel resources
- ◆ installed client base
- ◆ source of capital

Vendor support services

- ◆ project management
- ◆ conversion
- ◆ training
- ◆ customization
- ◆ documentation

In emerging technologies such as imaging, few vendors can claim 10 or more years of experience. However, if you ask a vendor how many imaging installations they have performed, their answer should not be: "Do you mean including this one?" A vendor's proposal should include a list of 5 to 10 customer installations with names, addresses and phone numbers. These should be installations in which the vendor had primary responsibility—not clients in which their services were incidental. Each reference should include a brief description of the customer's business and the imaging application installed. Ideally, these references should parallel your own industry and imaging interests.

Closely related to vendor experience and stability is the question of vendor support. The most basic element of vendor support is the project-management plan, which defines the responsibilities of the buyer and the seller during each phase of the project. Vendors should be given flexibility in developing a plan; however, any constraints (such as a desired completion date) should be noted. Although vendors should be free to define the content of the plan, the buyer should define its format. The RFP should include a request for the following:

♦ name and resume of the project leader

♦ experience level of support staff (with names and resumes, if possible)

♦ project deliverables and due dates

♦ quality-assurance checkpoints

♦ responsibilities and roles in the buyer's organization

♦ frequency of written project updates (usually weekly or biweekly)

Your IS group may have a formal project-management methodology that is authorized internally, such as Method One, Navigator, or The Guide. You might provide a copy of this methodology to the vendor to follow when developing the proposed project plan.

Within the project plan, vendor responsibility concerning issues of software customization and integration, site preparation, conversion, training and documentation should be defined. For the vendor to provide an accurate cost estimate, you must decide (on a preliminary basis, at least) what tasks your organization can perform. Software can often be customized in-house, but carefully consider whether your IS staff can develop the required expertise quickly enough to ensure a successful project. Software customization is performed in-house by only 25 percent of imaging buyers.[6]

Few imaging applications are turnkey, but many vendors offer prototypes of popular applications, such as accounts-receivable and accounts-payable. Development effort can be reduced significantly if a suitable prototype exists. You may want to solicit two quotes: a "turnkey" cost, assuming that the vendor does the software customization, and an off-the-shelf cost, assuming that you perform the customization. If you describe the skill set of your IS staff in the RFP, the vendor may advise you on a build-or-buy decision. Alternately, you can ask the vendor to provide job descriptions and skill levels for development staff.

Related to software customization is the issue of integration with existing applications. In your accounts-receivable department, for example, an application screen with payment history may be overlaid by images of a customer's invoices. Best results often are achieved by rethinking and redesigning applications rather than by simply grafting on the images. A heavily integrated imaging solution may require increased involvement from your IS staff familiar with those applications.

Responsibility for system installation, documentation and training usually falls to the party who designed the software. If a vendor is responsible for installation, the project plan should include a formal sign-off for system acceptance. Some payment holdback should occur until this quality-assurance checkpoint is reached. If the vendor is responsible for system documentation, the RFP should request sample documentation from prior engagements. The format and detail of the content should be defined for the vendor.

---

6. Leonard Yencharis, "What Large-Scale End-Users Want—And What They Get," *Advanced Imaging*, September 1991: 14.

Vendors may provide a comprehensive training program, or at least support the buyer in the development of program materials and formats. Requested details for training programs should include their locations, frequency, duration and content. Vendors should note any prerequisites for training, such as familiarity with a specific database system or UNIX. Classes may be provided for system managers, operators, end-users and management.

Pricing Section

The pricing section provides a structured format for vendors to follow when presenting the cost of their solutions. Listed elements will include:

- hardware
- application software
- system software
- communications hardware and software
- installation
- training
- maintenance
- documentation
- project-management resources

You should ask that prices be unbundled to show the specific cost of all products and services. This will enable you to compare "apples-to-apples" when weighing one proposal against another. Hardware should be broken down to the lowest component level—servers, storage, displays, scanners and printers. In the pricing of services, non-recurring costs, such as installation and conversion, should be separated from recurring costs, such as training and maintenance. "There may or may not be room to negotiate on a total cost of $2.4 million," according to Bud Porter-Roth. He notes that vendors often load programming expense into the cost of workstations, "which you have to buy."[7] However, if one vendor's workstations are $10,000 and another's are $20,000, buyers should identify the cause of the difference. Not all components, of course, can be unbundled. Some hardware and software (such as jukeboxes with proprietary software) may not be separable.

Contracts and Licensing Agreements

Your RFP should include a request for the purchase agreement, warranty, maintenance agreement and any performance bonds. An important question is the issue of licensing for software for the application, system and dataserver, workstations, and communication network. Will you be responsible for upgrades, or will these be a continuing service from the vendor? This section may also include reference to non-disclosure agreements. For competitive reasons, you may not wish an announcement to the press when a contract is executed. This stipulation can be included in the contract.

Quality RFPs Mean Successful Projects

The Request for Proposal is the concrete result of your task force's review and planning effort up to the point of purchase. Because successfully implemented imaging systems demand a willingness to reconsider how business is conducted, RFP development carries the weight of any other strategic-planning process. In other words, it can be expected to start with much enthusiasm that may disintegrate into second thoughts and dis-

---

7.   Porter-Roth, interview.

illusionment before the RFP jells into a clear definition of purpose and requirements.

A Request for Proposal, however, is worth the effort. The document sets a controlled, professional tone for the purchasing process and—when done skillfully—creates an agreement between buyer and seller on a relationship that may continue for a decade or more. It clarifies an understanding of needs not only for the task force and vendor, but for senior management whose support will be crucial as implementation proceeds. Stated most simply, the RFP raises the level and quality of communication and makes project success more likely.

The Request for Information

Before a proposed business solution is presented in an RFP, it may be wise to solicit vendor comments or the advice of a consultant experienced in imaging RFPs. An RFP under development can be stamped "preliminary draft" and comments can be solicited from two or more vendors. A preliminary RFP is sometimes called a Request for Information, or RFI. While no vendor should be allowed to skew an RFP toward its own technology bias, a vendor may help identify the following issues:

The RFP is too vague to elicit an effective response.

RFPs often are weak with regard to issues of document sizes, volumes and processing tasks. Unless described properly, familiar processing tasks within an organization may be incomprehensible to a third party. The Warnier-Orr diagrams discussed in Chapter 2 are an effective communication tool.

The RFP is so precise that superior solutions are overlooked.

In general, RFP requirements should be defined in functional rather than technical terms. An RFP can be too precise if it dictates a specific technical solution. It is essential, however, for a vendor to understand the technical strategy and environment of the organization. The vendor's proposal should be compatible with this environment (for example, support a technical strategy such as IBM's System Application Architecture), unless it would limit effective solutions.

Requested solutions that are technically infeasible.

A solution that demanded 100 percent accuracy in the recognition of free-form handwriting, for example, is not feasible with today's optical character recognition (OCR) technology.

Potential solutions overlooked because the user is unaware of imaging capabilities.

Conversely, opportunities may be overlooked that could be supported with today's technology. Full-text search, which allows indexing by any word in an imaged document, is feasible. This capability is especially valuable in support of legal research and litigation.

4

**Building a
Business Case**

*"If a picture is worth a thousand words, how much is a
document image worth?"*
—Peter G. Kane, Barclay's Bank[1]

The phrase *cost-justification* is often used to describe the financial
analysis of an investment in technology. Traditional cost-justifi-
cations are frequently static, one-time reports that gather dust
following system implementation. Once approved by a CFO, no
further evaluation of the investment occurs. Technology, however, must be managed dynamically for the best results to be
achieved, and this is especially true with imaging technology.

   As with general office automation, the use of an imaging
system is often voluntary. An employee may choose to use it or
not use it. This type of automation is called an *elective system*. If
the user is poorly trained or intimidated, or if the system is badly

---

1.   Peter G. Kane, "Alchemy, Synergy and Other Considerations in
     Cost-Justifying Imaging Systems," *Inform*, May 1991: 24.

designed, the user may simply circumvent the technology and continue a manual, paper-based process. While "cost-justified" in theory, the organization now bears the cost of both a manual and automated process.

An investment in imaging represents, in a very real sense, a double-or-nothing "bet" for your organization. Some organizations may even overlay a weak office automation system with a poorly designed imaging solution. An organization can then find itself with a triplication of effort—paper forms, electronic forms and images. When no performance measures exist, automation often degrades rather than improves productivity.

An imaging investment deserves a detailed *business case* that defines the role of imaging in the organization. This business case should include the following:

- ♦ the proposed business solution

- ♦ operational and strategic benefits

- ♦ direct and indirect costs

- ♦ a cost/benefit comparison

- ♦ a plan to monitor performance

This business case will link and clarify several issues identified in previous chapters, as well as issues addressed in future chapters:

- ♦ A proposed business solution was developed by your task force for the RFP, as described in Chapter 3.

- ♦ Operational and strategic benefits are a more detailed response to a question posed in Chapter 1: "Why will our organization benefit from document imaging?"

♦ Cost projections are drawn from vendor responses to the RFP. The development of cost projections will be described in further detail in Chapter 5.

♦ Performance measurement is the culmination of the project—a tool to guarantee that corporate goals are met. Performance evaluation and fine-tuning are discussed in Chapter 8.

Development of a *business case* thus began early in your project. As information becomes available throughout the project, it refines your knowledge of the opportunity. A business case should always be summarized before a vendor purchase, but it may be updated throughout implementation. A business case for imaging can provide the following benefits:

♦ The imaging investment is explicitly linked to corporate strategy.

♦ The quantification of benefits obliges management to identify the way in which specific results are achieved.

♦ The quantification of costs for varied vendor solutions can help identify an optimal benefit/cost balance.

♦ The business case becomes a performance yardstick to ensure that anticipated benefits are achieved.

A Proposed Business Solution

The business case (or cost-justification) begins with the proposed business solution—a concise statement of the problem, with an explanation of how document imaging can address the problem. Alternately, the business solution may be phrased as an opportunity—how imaging might improve a situation which is good but could be better. A business solution includes the following elements: a problem or opportunity statement; a description of how imaging can be applied; and a brief statement of anticipated benefits.

The problem statement may be several sentences or several pages, but the description of the imaging solution will typically be several pages in length. In systems-design parlance, this is called the conceptual design for the system. It conveys the scope and functionality of the planned system, and will be referred to (and perhaps refined) during the development process.

As noted in Chapter 1, the benefits statement can and should be brief: "Our goal with imaging is to measurably improve customer service while reducing servicing costs by 20 percent." A concise vision of imaging's purpose will serve as a touchstone throughout your project.

Strategic Benefits

Topics introduced in your proposed business solution are expanded on throughout the business case, which should also include a section addressing strategic benefits.

The strategic benefits of imaging typically relate to the timeliness and quality of information. Time has been called "the calculus of business." What is the relative value of having information in five seconds, five minutes or five hours? How much more profitable is a commodities trader if he or she has data five seconds prior to a competitor? When a pharmaceutical firm provides test data to the FDA, does it matter if this information is retrieved in five minutes or five hours? Each business must answer these questions for itself. (Figure 4.1 illustrates the time value of data for one user.)

Closely related to timeliness is the issue of quality or accuracy. With imaging, retrieval accuracy is measured by two criteria: the *recall ratio* and the *precision ratio*. The recall ratio is the

Figure 4.1 Time Value of Data

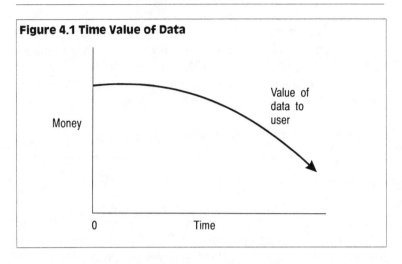

number of relevant documents retrieved, divided by the total number of relevant documents in the database. The precision ratio is the number of relevant documents retrieved, divided by the total number of documents retrieved. Assuming that all documents have been properly scanned and indexed, retrieval accuracy with imaging (both precision and recall) is always 100 percent—the document simply cannot be lost unless the optical disk is destroyed. What is the cost of a lost document? Again, each business must answer this question for itself.

In addition to providing information faster, and with a greater degree of quality, imaging offers data previously un-available. Paper never "informs" you of its status. Unless a document is in a file with a manual "processing checklist," it may be impossible to determine who has reviewed it or what actions have been taken. By itself, a document is merely a "snapshot" at one point in time—it usually conveys little or nothing concerning the flow of events. An imaging system, however, not only pro-vides the document but communicates its status—when it was received, who has reviewed it, what actions have been taken, and what deadlines must be met. It places the document in the

context of pertinent business events. The value of this information is often difficult to measure—it simply may prevent a phone
call, or it may ensure a crucial delivery date is met.

The strategic value of imaging potentially affects four
groups of stakeholders in your organization:

+ customers (clients, constituents)

+ financial investors (shareholders, taxpayers, lenders)

+ employees

+ interested third-parties (regulators, auditors, donors)

Potential strategic benefits resulting from imaging technology are addressed in the following pages by group.

Customer Value

In *Total Customer Service*, author William Davidow describes an
evolution of competitive strategy that occurs as markets mature.[2]
As a means of differentiation, there is an inevitable transition in
competitive strategy from features to cost, then to quality, and
finally to service. As Davidow explains:

> "With new products and services, competitors focus
> on features such as horsepower or direct dialing. Com-
> panies battle each other with 'specsmanship.' But as
> competitors swarm in, generally matching each other
> feature for feature, the skirmishing shifts to cost. In
> time though, customers get 'spoiled.' They come to
> expect competitive features, prices, and quality. The
> [final] battleground is service to the customer."

Today, service is a cornerstone of competitive strategy for
many businesses, whether they are manufacturers, retailers,

---

2. William H. Davidow and Bro Uttal, *Total Customer Service* (New York:
   Harper & Row, 1989).

service companies or distributors. Recognizing a trend, organizations in the public sector—health-care providers, government agencies and not-for-profits—have launched similar efforts toward responsiveness.

Customer service is largely a matter of perception. What may have been perceived as excellent banking service a decade ago, such as a two-minute wait for a teller, may be perceived as inferior service today by customers accustomed to instantaneous service from automated teller machines. Companies that offer service guarantees are intentionally "raising the stakes"—both for themselves and for their competitors. Installation of effectively designed imaging solutions can also raise the standards and the expectations within an industry. Potential measures of customer service include:

- order- or service-turnaround time
- number of complaints
- problem-resolution time
- customer-retention ratio
- revenue per customer
- pricing flexibility

Through imaging, processing time is reduced for service events such as approval of a loan application, resolution of an insurance claim, or response to a question on the redemption of a mutual fund. If the imaging system is properly designed, all documents needed to fulfill an order, answer an inquiry or respond to a complaint are instantly available on-line—enhancing both speed and accuracy.

Measures of service quality include average and maximum transaction times, and the percentage of "call backs" required to customers, when additional research is needed. The number of complaints received and the time required to resolve those complaints are also important measures of service quality. Imaging systems can be combined with service-tracking automation to monitor types of problems and resolution times.

Customer satisfaction can be measured indirectly through customer-retention ratios, cross-sell ratios and revenue/customer ratios. The customer-retention ratio is defined as the percentage of last year's customers retained this year as active customers. The cross-sell ratio, suitable in industries such as insurance, finance and retailing, is the number of products sold per customer. Revenue per customer also can serve as an indirect measure of customer satisfaction, as does the capability to charge a premium price for a product because of high service quality. With the assistance of your marketing staff and financial analysts, it is possible to translate service objectives into projected revenue gains. If a service goal is to reduce customer run-off by 25 percent, the financial value of these retained accounts can be estimated. Measures can be applied before and after system implementation to ensure the achievement of desired results. (See example analysis, Figure 4.2.)

A business can be differentiated through service; service, in turn, can be differentiated through imaging. However, unlike simpler, "shrink-wrapped" technologies, such as automated teller machines, the capability to improve service with imaging depends largely on the creativity applied in its conception and its design.

Investor Value

Value for the investor is closely correlated to value for the customer; when the customer is satisfied and spends money, the investor usually profits. Investors obtain additional benefits, however, through reduced operational expense combined with faster speed-to-market for new services or service improve-

Figure 4.2 Valuing Customer Satisfaction

	Pre-Imaging	Post-Imaging	Net Change
Service Criteria			
Error rate/1,000 transactions	43	12	(31)
Problem resolution time for walk-in service (in-minutes)	12.9	7.5	(5.4)
Problem resolution time for phone-in service (in minutes)	10.2	5.3	(4.9)
Impact on Customer Base			
Accounts per customer (cross-sell ratio)	1.2	1.5	.3
Customer turnover rate/year	32%	26%	(6%)
Customer retention period (in years)	3.7	4.1	.4
Impact on Profitability*			
Advertising expense/year (to replace lost customers)	$275,000	$203,000	($72,000)
Net income per customer/year (based on additional .3 accounts)	$570	$712	$142
Total net income per customer (over life of the relationship)	$2,109	$2,921	$812

*assumes net income of $475 annually per account

ments. The time required to conceive, design, prototype and manufacture products is greatly reduced through technologies such as CAD (computer-aided design) and CAM (computer-aided manufacturing). Benefits analagous to those achieved by manufacturers through CAD/CAM are provided through imaging to paper-intensive service areas in any firm.

In most industries, speed-to-market is closely related to market share and profitability. Market share and profitability are usually achieved by dominating an emerging market quickly. Imaging can help an organization, in the words of management consultant Peter F. Drucker, "be the first with the most." From an investor's perspective, the long-term success of imaging as a strategic technology will ultimately be measured in earnings per share and price per share.

Employee Value

Abraham Maslow's "hierarchy of needs" has served for several decades as a model to explain human behavior in the workplace. As the most basic human needs are addressed, such as the desire for security, they cease to be motivators. The employee looks to a new plateau for personal fulfillment. The highest plateau is referred to as *self-actualization,* and describes the desire of most employees to find satisfaction through meaningful and relevant work.

Maslow's theory is the foundation for concepts such as *job enlargement* and *job enrichment* where tasks are redesigned to provide workers with a sense of continuity and completion. The movement toward job enrichment is a reaction against specialization so extreme the worker is alienated. An employee whose life's work is "tightening the right bolt on the rear bumper" is probably unmotivated and adversarial, however well-paid the person. In the auto industry, job enrichment was successfully pioneered by Volvo and several Japanese automakers in response to this problem. Workers are responsible for all aspects of production, from scheduling to quality control.

Imaging presents an opportunity in the roles of workers for either regression or advancement. If poorly designed, the imaging system can create a sterile, monotonous and unrelenting work experience—a computer equivalent of the Little Tramp caught by the assembly line in Charlie Chaplin's movie, *Modern Times*. When properly designed, the imaging system should eliminate the routine frustrations of file hunt-and-search while creating an environment at least as fulfilling as its paper-based predecessor.

Third-Party Value

Interested third-parties can include directors, auditors, regulators, industry analysts or donors. They usually have an official interest in the organization, perhaps mandated by law or defined by professional codes. In general, they are concerned with the integrity of specific activities—the testing of pharmaceuticals, the preparation of financial records, the disposition of assets. Their work often requires sampling of records to determine compliance with regulations or concepts, such as GAAP (generally accepted accounting principles). An imaging system designed with some forethought can greatly facilitate these efforts. The cost of an audit, for example, may be significantly reduced if documents supporting financial transactions can be efficiently sampled and retrieved through the system. Imaging will likely elevate the importance of EDP auditors with technical expertise.

Organizations under stringent federal or state regulation, such as pharmaceutical firms, may stand to gain the most through imaging. If a drug company cannot document compliance with FDA guidelines, for example, approval to market a drug can be denied or withdrawn. The cost of missing documents may be simply unquantifiable.

Operational Benefits

The business case should also address potential operational benefits. Operational benefits include any direct or indirect savings achieved through imaging. *Direct savings* are reductions in the cost of handling or storing paper-based documents and files. *Indirect savings* are reductions in the cost of using these documents or files.

Direct Savings

Direct savings are achieved through an elimination or reduction in personnel, materials, equipment and miscellaneous processing expenses. Although the design of an imaging system may include tasks added to improve service quality, there is almost always a significant net gain in productivity. This productivity gain may result in the elimination of some positions and the displacement of other staff into new roles. Staff reductions may be achieved through layoffs or natural attrition, depending on issues such as your corporate philosophy or existing labor contracts. In some high-growth organizations, new volumes surpass any productivity gains, so existing staff will simply process these additional volumes more efficiently. If a reduction in staff occurs, the savings is called a *cost reduction*. If additional volumes are handled more productively, the savings is called a *cost avoidance*. In either event, the true net savings in personnel must be identified based on the replacement of your paper-based system with an imaging solution.

Figure 4.3, developed by cost-justification expert Ray Abi, presents estimated clerical time required with an image-based system. More precise figures might be developed internally through time studies and refined as the imaging system is installed. Savings in physical resources with imaging also are considerable, and can include the following:

Figure 4.3 Estimated Clerical Production Time, With and Without Imaging (in Minutes)

Clerical Functions	Tasks Included	Pre-Imaging	Post-Imaging	Net Change
Storage	Review, index and store	4	3	(1)
Retrieval	Search, retrieve and printing	Simple . . . . 3 Complex . . .5	1	(2) to (4)
Processing	Actual processing of document	Varies with type of activity. Estimate needs to be developed by the analyst.	Varies with type of activity. Estimate needs to be developed by the analyst.	varies
Searching misfiled documents	Document search	30	None	(30)
Copying	Copy document, return original document to files, and mail copy	5–7	5–7 (Same as manual system except volume is substantially reduced)	None
Microfiche/Microfilm	Setup, filming, storage	2	2 (Same as manual system except volume is reduced)	None
Telephone	Dialing, talking	5–15	3–10	(2) to (5)
Mailing	Stuffing and addressing	1	1 (Same as manual system except volume is reduced)	None
Error Correction	Document review, calling other people, document correction	15	5–10 (Depends on the type of error)	(5) to (10)

Source: Ray Abi. *Justifying Image Processing Systems*. Stamford, CT: Unitech International Corp., 1991: 9.

- file cabinets and file space
- paper (forms, documents, records)
- storage space for records
- micrographic production equipment and supplies
- photocopiers and supplies

The savings in floor space for both office file cabinets and off-site records storage should be identified. A single 12-inch optical disk can easily store the images of 20,000 pages of text, or the equivalent of four filing cabinets. A jukebox holding 100 optical disks can store the equivalent of 400 filing cabinets or 2 million pages of text. Elimination of off-site records may generate a direct cost reduction: leased space can be sub-leased and owned buildings can be leased or sold. If documents are stored by a records-management firm, contract costs can be easily identified. Space savings in an office usually translate into cost avoidance: the need to lease or purchase new office space is delayed as staff expands.

Micrographic production equipment and supplies may be affected by an imaging system. Imaging systems intended for archival storage-and-retrieval can be partially cost-justified based on a reduction in micrographic production expense. The cost of acquiring and maintaining microfiche readers should not be overlooked. The net savings in space should also be identified. Imaging systems will require PC workstations, but stand-alone microfiche readers may be eliminated entirely.

Imaging also can reduce photocopying costs—a large but often hidden expense. The annual cost of maintaining a copier can range from $5,000 to over $50,000, depending on the type of machine and work volumes processed. An extremely thorough analysis of opportunities for reduced photocopying expense was performed by Robert Zeek for Pfizer Corporation. "A substantial cost-justification can be made," according to Zeek, "without any reference to the so-called 'paperless office'."[3]

Zeek's division in Pfizer generated 8 million photocopies per year, using 20 laser printers and 30 copiers. The cost per page of documents produced via copiers was determined to be $.0588. The cost per page of documents produced via laser printer was $.0283. Assuming that the 30 existing copiers were replaced by 30 laser printers, annual projected savings would be $244,000. Only $30,000 of this figure is from "soft" labor costs. Supporting detail is presented in Figure 4.4.

Imaging may also eliminate miscellaneous processing costs: mail, overnight shipments, or phone calls concerning lost documents. These costs can usually be identified in a review of departmental expense records.

Indirect Savings

Indirect savings are cost reductions or productivity gains achieved by end-users. In some archival applications, there may be no distinction between the storage or retrieval of documents and the processing of those documents. With work-in-process systems, however, some professional review or processing of the documents will occur after these documents have been retrieved by clerical staff. This review or processing time may be reduced by 50 percent or more through an imaging solution, especially if the system includes OCR and workflow automation that allow the following:

♦ incomplete documents to be returned for completion

♦ documents to be indexed by OCR

♦ relevant documents to be retrieved, when an integrated application is accessed

3. Robert Zeek, "Digital Document Image Automation: The Case for 'Paperless' Files," *Optical Information Systems*, January-February 1988: 13.

Figure 4.4 Direct Savings — Photocopier Expense

Copiers (30 at approximately $600k):

1. Depreciation expense	.0113	
2. Maintenance costs at 10 percent	.0150	
3. Supplies (toner, cartridges, parts, etc.)	.0200	
4. Paper.	.0050	
5. Labor at copier	.0075	
Total cost per page		.0588

If lasers replace all copiers (30 at about $240k):

1. Depreciation	.0023	
2. Maintenance costs at 10 percent	.0030	
3. Supplies (toner, cartridges, parts, etc.)	.0180	
4. Paper.	.0050	
5. Labor at printer	.0000	
Total cost per page		.0283
Difference per page		.0305

$$\frac{\times\ 8\ \text{million copies/year}}{\$244{,}000\ \text{per year}}$$

Source: Robert Zeek. "Digital Document Image Automation: The Case for 'Paperless' Files." *Optical Information Systems.* January – February 1988: 12–13.

♦ documents to be automatically routed to the next re-
viewer

♦ work to be prioritized, based on deadlines or dollar value

The savings achieved will vary for each application. Some
new tasks may be engineered into your process to improve
quality control or customer service, while many existing tasks
will be eliminated. Savings can be estimated through time studies, but one effective and simple alternative is a review of the
procedural diagrams (or Warnier-Orr charts) constructed for the
manual process. Before-and-after diagrams can be compared to
identify the total number of original manual tasks, and the
number of tasks eliminated or automated. This approach assumes that the average processing time per task for both the
eliminated tasks and for the remaining tasks are approximately
equal. Because of this statistical assumption, the method is most
accurate when reviewing a larger process. If manual processing
originally required 500 tasks and 240 of those tasks were eliminated or automated through imaging, productivity savings
should approximate 240/500, or 48 percent. This approach has
the distinct advantage of employing research probably already
developed: before-and-after procedural diagrams for the process to be automated (similar to those in Figure 2.3).

Direct and Indirect Costs

Your business case should also include a discussion of direct and
indirect costs. Direct and indirect costs of imaging are addressed
in Chapter 5, in the context of vendor evaluation. However, just
as there are strategic benefits with imaging, there are also some
costs that are *intangible*. One *intangible cost* of an imaging solution
is the opportunity cost of management's time and effort. The
opportunity cost is the value this time might have offerred had
it been committed to another endeavor.

Another intangible cost is the disruption associated with change. A conversion to any new system or procedure carries an implicit cost as the new approach is being mastered. These costs are minimized through training and effective change management. An imaging conversion may be viewed positively, however, as a catalyst to facilitate organization-wide change. The restructuring needed for an image-enhanced process can be accompanied by worthwhile changes in job descriptions, incentive plans and performance measures.[4] Ideally, the result is job enrichment and improved employee morale, as staff assume responsibility not simply for the quality of the *product*, but also for the quality of the *process*.

Financial Analysis

The benefits and costs of the imaging system must be compared in order to document that an expenditure is justified. In for-profit organizations, any expenditure must ultimately contribute to the generation of increased income. The cost of an imaging system should therefore be viewed as an investment, rather than as an expense. In any business, potential investment opportunities constantly compete for resources. Because investment capital is always limited, only the best opportunities are funded. To receive funding, it is not sufficient to simply show a positive return on an investment (or, worse yet, to simply show cost recovery). It is necessary to establish that an investment is superior to the other investment alternatives available to senior management.

The most widely used financial measures are *return on investment* (ROI) and the *payback period*. An ROI calculation begins with an identification of the cost of the project, year by year, over a specified time period. This cost is netted against the direct

---

4.   Edwin D. McDonell, "A Catalytic Systems Conversion," *Information Strategy: The Executives' Journal,* Summer 1990: 16-19.

and indirect savings, to establish a standardized rate of return on the investment. This return on investment, or ROI, then serves as a universal benchmark to compare investment alternatives. A voice-mail project yielding an ROI of 15 percent can be weighed, for example, against an investment in advertising yielding 12 percent.

With the possible exception of government-insured bank deposits or Treasury bills, no investment is risk-free. Every investment opportunity has its own unique risks. And as all experienced investors know, investments with greater potential risks must offer greater potential returns if they are to be considered. The payback period is one means of measuring risk. The payback period, stated in years or months, is an estimate of how much time is required to recover the original outlay of funds (that is, the principal of the investment). When two investments promise an equivalent ROI, the option with the shorter payback may present less risk, since the funds are committed for a shorter period.

More difficult to estimate than the payback period is the probability of achieving the anticipated ROI. A risk-free investment, such as an insured bank deposit, has a probability of nearly 100 percent—unless the U.S. government fails, both principal and interest will be paid. Senior management is sometimes accused of short-sightedness when considering investments in automation. While some members of management no doubt are short-sighted, most are grappling with the very legitimate issue of investment risk. "I'm as concerned about return *of* my investment," humorist Will Rogers once quipped, "as I am about return *on* my investment."

Those challenged to justify an investment in imaging should consider the crucial aspects of a funding decision. As the business case is quantified, it should help shape the timing and scope of an imaging investment. Rather than the static "all or nothing" analysis often employed with automation, the flexibility of imaging allows a more dynamic approach. Within the task force, your financial analysts, IS staff and users can collaborate to refine an imaging plan that offers the best possible investment return. Issues to consider include *self-funding, performance measures* and *quality assurance.*

Self-Funding

The goal of self-funding is to manage the scope and timing of expenditures so that investment outlays occur gradually. The payback from your early investment will help fund your future investment. With self-funding, first priority is reserved for high-payback opportunities with lower risks or lower costs. Simple archival solutions may be pursued first, followed by more complex integrated imaging applications. Advanced OCR technology or expert systems may be the final stage of investment. A self-funded project may even begin with a work-redesign study to simplify manual procedures and improve the use of your existing automation. In his book, *In Search of Excellence*, Tom Peters calls this approach, "the technology of the obvious."[5]

Performance Measures

Performance measures are closely related to the concept of self-funding because quantitative data is needed to determine payback. Performance measures should be applied to ensure that stated goals are achieved. A single performance objective that can only be measured at the end of a five-year investment period is inadequate. Interim measures, such as the ratio of customer inquiries answered per hour, should be applied to verify that five-year objectives are met. Combined with the concept of self-funding, senior management should greatly appreciate periodic

---

5.   Thomas J. Peters, *In Search of Excellence: Lessons from America's Best-Run Companies* (New York: Warner Books, 1982).

performance measures that demonstrate phase-by-phase funding objectives are being met.

Quality Assurance

Broader in scope than performance measures, quality-assurance checkpoints help you ensure that project deliverables meet established due dates and match your defined needs. From management's perspective, these checkpoints also serve to reduce the business risk associated with an imaging investment. Investment risk is reduced, obviously, through careful attention to the following managerial aspects of an imaging project:

◆ appointment of an experienced task force leader

◆ selection of capable task force members

◆ an innovative systems design that supports service goals

◆ detailed evaluation of all suitable vendors

◆ involvement, support and ownership by end-users

◆ detailed planning for training and implementation

A task force leader who strives to minimize business risk will be highly valued by senior management. Skilled IS professionals recognize that, in a business setting, investment risk and technology management are interrelated topics.

Net-Tangible-Cost Model

The financial measures previously described relate only to tangible or quantifiable costs and benefits. If a dollar value cannot be assigned to a benefit, it cannot be included in any traditional measure of investment performance. However, your qualitative or intangible benefits from imaging may be very significant. The *net-tangible-cost model* supports decisions that include both quantitative and qualitative issues (see Figure 4.5). Following this approach, tangible (hard-dollar) costs are compared to tangible (hard-dollar) benefits. The resulting figure may show a *net-tangible-cost* or a *net-tangible-benefit* for the project. If the figure shows a net-tangible-benefit, then the project is clearly justified based solely on quantifiable data.

Sometimes, however, there is a net-tangible-cost associated with the project—that is, a hard-dollar cost that is not fully matched by hard-dollar benefits. In that event, the net-tangible-cost can be compared to the intangible or qualitative benefits associated with imaging. The strength of this approach is that it allows intangible benefits to be weighed against the project costs not fully covered by tangible savings, thus facilitating an intuitive judgment. In addition to traditional measures used to justify an automation investment, it is useful to translate these costs into a format such as the *net-tangible-cost model* that can support an intuitive decision.

Finally, it should be noted that the terms "tangible" and "intangible" represent two ends of a continuum, rather than two

---

Figure 4.5 Net-Tangible-Cost Model

Tangible Benefit (direct and indirect)	$479,125
Tangible Cost	−297,750
Net Tangible Benefit (five-year present value)	$181,375

Intangible Benefits:
- Improved customer service
- Improved regulatory compliance
- Faster product modifications
- Improved morale

Source: Edwin D. McDonell, "Corporate Reengineering that Follows the Design of Document Imaging." *Information Strategy: The Executive's Journal.* Fall 1991: 8.

distinct categories. Many benefits that could be quantified (such as increased revenue from improved service) remain intangible because of a lack of analysis. Figure 4.6 presents potential benefits with imaging, and the estimated difficulty of quantification.

Intuitive Analysis

Another means of intuitive analysis is to recompute your figures either on a per-person or per-transaction basis. It is difficult to comprehend either costs or benefits stated in millions of dollars, as they often must be in large organizations. It is sometimes more meaningful to say: "Imaging will increase the cost of processing a customer inquiry by $.78, but will reduce the time spent from an average of 2.3 minutes to 1.4 minutes." This type of analysis enables management to assess an imaging decision on an intuitive level. Understanding the value of a customer's time and the value of a customer's account relationship, management can logically judge the tradeoff of time and expense.

The Cost of Inaction

Imaging will almost certainly demand a significant capital investment. Per workstation costs typically exceed $20,000. An entry-level client/server system will cost at least $100,000, while enterprise-wide solutions cost millions of dollars. Such expenditures will be intently scrutinized by senior management and the board of directors. And even when the value of imaging is recognized in theory, other worthwhile projects are competing for the same resources.

Your task force may wish to consider the cost to your organization if imaging is *not* pursued. The potential benefits of imaging were estimated in your business case, but what is the cost of inaction or delay? What will happen when your competitors speed their product development, enhance their customer service, or better control their costs through imaging?

Figure 4.6 Intangible Benefits of Imaging

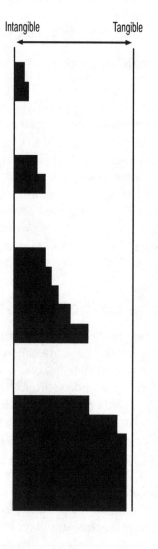

Intangible   Tangible

Service Management
- Fewer Errors
- Faster Service

Risk Management
- Increased Document Integrity
- Increased Document Security

Revenue Management
- New Products
- Faster Speed-to-Market
- Increased Attraction of New Customers
- Increased Retention of Current Customers
- Increased Sales to Current Customers

Expense Management
- Reduced Floorspace Requirements
- Reduced Document Storage Costs
- Reduced Document Supplies Costs
- Reduced Document Distribution Costs
- Reduced Training Costs
- Increased Productivity

To measure relative performance, organizations often rely on market share, and per-unit production or transaction costs. Because there is usually a long-term correlation between the quality of customer service and market share, the cost of inferior service might be quantified as a percentage decline in market share. And if there is a correlation between profitability and expense control (which is usually the case), the cost of inaction will be narrower profit margins—as your imaging-enabled competitors manufacture products or deliver services more efficiently.

While imaging is optional today for many organizations, in 10 years or less it will be mandatory for most paper-intensive companies. When the majority of organizations in a business sector are image-enabled, it will simply be a *technological imperative* for the remainder. For organizations that delay action, the cost of this lost opportunity may be measured all too easily in the long-term, as their revenue, profitability and stock prices decline.

5

Vendor Evaluation

"Will all documents as we know them today disappear
in a fluff of bits and bytes as we have so long been told?
Don't count on it."
—Robert J. Kalthoff, President, Kalthoff & Associates[1]

After a Request for Proposal has been prepared, vendor candidates must be identified to receive the RFP. Identification of vendor candidates and the evaluation of vendor capabilities are perhaps the two most crucial aspects of an imaging decision. It may be impossible to recover from an error, because it will not usually be evident that a mistake has been made. Through lack of research, an organization may neglect a leading vendor can-

---

1. Robert J. Kalthoff, "Large Scale Document Automation: The Systems
    Integration Issues," *Videodisc and Optical Disk*, November-December 1985:
    453.

didate. And if a poor evaluation methodology is applied, the best candidate may still be overlooked.

The risks inherent in any automation decision are perhaps even greater with an imaging investment. As described by consultant Mary Ann O'Connor, "Image systems are 'closed' systems. Once you make a commitment to a vendor, you are required to implement their complete hardware/software solution."[2] This may be a slight overstatement (for example, UNIX-based client/server systems are hardware independent), but the user's commitment to a vendor is nonetheless significant. Steps to be performed include the following:

1. Identify vendor candidates.
2. Issue a Request for Proposal.
3. Communicate with vendors concerning the RFP.
4. Evaluate proposed vendor solutions.
5. Select a vendor solution.
6. Negotiate and finalize a contract.

Identify Vendor Candidates

Once completed, the Request for Proposal can be used to narrow the field of vendor candidates. Issuing an RFP to a dozen vendors indiscriminately is not the best approach—it generates more work for you and is unfair to the vendors. If suitable vendors were not identified through a Request for Information, the RFP can help you identify four or five leading candidates.

When selecting vendor candidates, consider the following:

♦ Unless a simple, archival application is sought, vendors of "electronic file cabinets" should not be considered.

---

2. Mary Ann O'Connor, "Applications of Optical Technology: Information Retrieval," *Document Image Automation*, May-June 1991: 168.

These types of solutions cannot be expanded enterprise-wide.

♦ The organization's primary hardware vendor should receive an RFP because they may offer proprietary solutions compatible with existing application systems.

♦ Vendors may be eliminated if they are incompatible with a mandated technology strategy. However, these judgments should be made carefully; there is benefit in allowing vendors to challenge the status quo.

As noted in Chapter 1, vendors may offer hardware only, software only, hardware and software, or may be independent integrators. To ensure a robust study, the list of vendor candidates may reflect competing strategic options: client/server versus mainframe; open systems versus proprietary systems; turnkey solutions versus in-house development.

Situations to Avoid

A good study can easily be compromised by the following errors:

♦ Rigging the study intentionally or unintentionally by including only one viable vendor solution in the field of candidates.

♦ Including several vendors offering identical solutions.

♦ Eliminating technically viable solutions for extraneous reasons.

As any marketer can attest, more than a few studies are rigged in advance by pitting a popular vendor (often the incumbent hardware provider) against recognizably weak alternatives. A vendor offering a comprehensive mainframe solution will not receive much competition from a vendor proposing a single workstation, archival system. A mainframe proposal may be second-best, however, when compared to solutions on a client/server platform. Several vendors also offer similar or even identical software, resold and relabeled for marketing purposes. When three or four vendors offer similar solutions, an evaluation study considering only those alternatives may be meaningless.

Finally, do not eliminate qualified candidates solely on the basis of financial viability. If a vendor with a superior solution fails, the system is usually acquired by a larger, better-capitalized firm. The acquirer will have a vested interest in meeting contractual obligations—your future maintenance fees are the major financial asset they purchased. Buyers can obtain reasonable financial protection through performance bonds, source-code escrow agreements, or third-party guarantees. Some larger integrators offer "viability guarantees" as part of their marketing strategy. GTE/Contel, for example, assumes responsibility for all integrated applications they install, even if the vendor should fail.

The *AIIM Buying Guide and Membership Directory*, published by the Association for Information and Image Management, offers a comprehensive list of vendors, integrators and consultants.[3] Further detail on hardware and software providers is available from Datapro. *Datapro Reports on Imaging Technology* contain detailed information on imaging technology, hardware and software vendors, and the industry.[4]

Datapro's most valuable service may be the charts profiling and comparing technical characteristics of most imaging solutions. In addition to including the usual "feeds and speeds" data, detail is provided on compatibility with various operating systems and network protocols. Monthly updates are provided to

3. Association for Information and Image Management, *Buying Guide and Membership Directory* (Silver Spring, Md.: AIIM, 1993).
4. *DataPro Reports of Document Imaging Systems: Index and Vendor Directory* (Delran, N.J.: McGraw-Hill, 1991).

ensure the data is current with vendor releases. Although the data provided by vendors to Datapro is not subject to verification, it is widely distributed and can serve as a check against representations made by overly enthusiastic marketers.

Vendor Communication

Vendors or integrators who will receive RFPs should be notified as soon as possible so that they can make resources available to respond to your document. A vendor will typically need four to six weeks to respond to a detailed RFP. As noted in Chapter 3, all communications to and from the vendors should be through a single liaison within your organization. This communication should be actively managed as vendors are preparing their proposals. To avoid misinformation or favoritism, a log of vendors' questions and your responses can be maintained. A bidders' conference may also be held (see Chapter 3).

A periodic, written update to all vendors should occur concerning any revision or expansion in RFP content. This is particularly important for government agencies that are publicly accountable for an unbiased study, but this discipline can save any organization much time later. The evaluation of RFPs is often delayed, when proposals based on inconsistent data must be reconciled. A few minutes spent communicating with vendors can save hours later in the evaluation phase.

Unless your business is a major financial opportunity for the vendor, its representatives may be reluctant to follow your requested proposal format. Vendors typically prefer to respond with their own "boilerplate" proposals. This is easier for them but will make your evaluation much more difficult because you will need to compare and reconcile different proposal formats. To ensure a suitable response, emphasize the fairness and logic of your evaluation criteria. The best imaging vendors usually relish a detailed comparison of features and price, while weaker vendors avoid this competition. If vendor candidates have been properly selected, they should all offer viable solutions, but the proposal process may prompt some vendors to disqualify themselves if they believe they cannot win. This saves time for all parties involved. Finally, the importance of active communication with vendors cannot be overestimated. If leading candidates do not prepare suitable proposals, the entire process must be repeated, setting your project back by a month or more.

Review of Proposals

Once received, vendor proposals will need to be reviewed. Having formal evaluation criteria ensures that proposals are judged according to their merits—rather than on the attractiveness of the proposal format or the friendliness of the vendor's marketers. Several possible scoring methods lend themselves to different stages of the RFP evaluation.

A binary approach (yes/no—does or does not meet criteria) can be used for the initial "cut" of responses. Some proposals may be eliminated based upon functional, technical or management concerns, such as the following:

♦ If a vendor cannot support OCR capability and it is a cornerstone of the desired solution.

♦ If no communications gateway exists to link the imaging system with existing applications.

♦ If a vendor cannot provide adequate or timely project leadership.

Ideally, vendors that cannot meet mandatory requirements are not issued RFPs or choose not to respond. However, late developments, such as a delayed product release, may quickly reduce the viability of a leading candidate. Keep in mind that vendor evaluation is sufficiently time-consuming without devoting your effort to obviously unsuitable candidates.

It is not unusual for some vendors to receive RFPs as a courtesy because they are the incumbent hardware vendor, for example, or have a special contact with a key executive. Vendors that insist on responding should be dropped from further evaluation once critical weaknesses are documented. However, vendors should not be dropped from evaluation based on price. Leading candidates may seek to charge a premium price, but cost issues can be addressed later, during contract negotiation.

Detailed Evaluation

After proposals are reduced to a "short list" of viable candidates, detailed evaluations can be made. Avoid the mistake of a decision driven by functionality or by price. Buyers considering only functionality may acquire more capability than can be cost-justified. Buyers considering only price can invest in solutions that fail to meet long-term needs. Further costs are incurred when the previous system is scrapped.

Most imaging needs fall between the extremes of a "Chevy solution" or a "Cadillac solution." The goal is the best value for the cost incurred, that is, the best possible return on your imaging investment. A structured methodology can apply this concept of value to your imaging decision. Figure 5.1 presents evaluation criteria and the points awarded to three vendor candidates. The benefits of each solution are measured by points awarded, based on the vendor's capabilities. The value of each vendor solution (reflecting their capabilities and cost) is shown in Figure 5.2. Points awarded for capabilities are plotted on the vertical axis of the graph. The cost of each solution over a five-year period is plotted on the horizontal axis. The candidate with the highest benefit/cost ratio offers the best value. In this example, Vendor A is the candidate that provides the best value, with a bene-

Figure 5.1 Vendor Evaluation Criteria

Technical Criteria	Potential Points	Vendor A	Vendor B	Vendor C
Image Quality/Resolution	5,000	3,200	3,500	2,700
Storage Media	5,000	3,376	3,729	1,276
Security/Recovery Features	2,500	1,542	1,762	789
Connectivity/ Interdependability	5,000	2,963	3,276	1,278
Communications Capability	5,000	3,724	3,732	1,183
Capture/Retrieval Speed	5,000	4,010	3,400	1,457
Software Feaures/Functions	10,000	5,930	7,039	2,887
Ease of use	10,000	6,279	6,428	2,394
Customization Capability	7,500	4,899	5,963	2,177
Integration With Existing Equipment/ Software	10,000	6,535	6,923	2,629
Expansion Capabilities Without a Conversion	5,000	3,265	3,519	1,444
Technical points	70,000	45,723	49,271	20,214

Figure 5.1 Vendor Evaluation Criteria (continued)

Management Criteria	Potential Points	Vendor A	Vendor B	Vendor C
Company History/ Longevity	2,000	829	1,510	570
Financial Performance	4,000	1,922	3,123	1,142
Installed Client Base	5,000	2,586	3,176	1,226
Source of Capital	3,000	1,293	2,701	865
Project Management	3,000	1,558	2,263	805
Conversion	2,000	961	1,486	629
Training	4,000	2,197	3,020	1,308
Customization	4,000	1,628	3,192	1,096
Documentation	3,000	1,442	2,182	927
Management points	30,000	14,416	22,653	8,568
Total -All points	100,000	60,139	71,924	28,782

Figure 5.2 Vendor Capabilities and 5-Year Net Present Value

	Functional Points	5-Year NPV Costs	Benefit/Cost Ratio
Vendor A	60,139	$1.28MM	.04698
Vendor B	71,294	$2.37MM	.03035
Vendor C	28,782	$1.26MM	.02284

fit/cost ratio of .04698. The graph line for Vendor A has the steepest slope and illustrates this concept visually (Figure 5.3).

The solution proposed by Vendor A does not offer the greatest level of functionality or the lowest cost. Vendor A does offer, however, the most functionality per dollar. Most vendor survey methods consider cost an evaluation criterion to be compared with system functionality. It is more appropriate to consider cost the denominator in the benefit/cost ratio, because it is a measure of value for the capabilities provided. In this example, it is interesting to note that the best business solution would have been overlooked had the primary criterion been either functionality or cost. In support of this methodology, the remainder of the chapter presents techniques for validating vendor capabilities, preparing cost projections and negotiating with vendors.

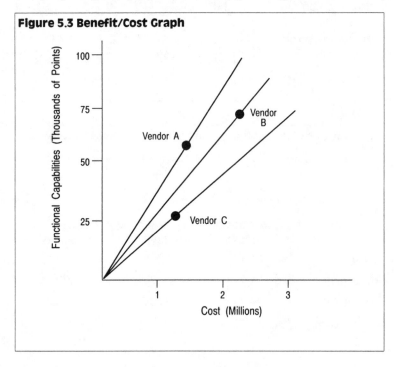

Figure 5.3 Benefit/Cost Graph

Evaluate Vendor Solutions

Vendor solutions are graded using a review of documentation, demonstrations, site visits and user surveys. Users and IS staff should verify functional and managerial capabilities, while IS staff verify technical performance. The time needed for evaluation may overwhelm users already preoccupied with daily work. User participation, however, is essential. Uninvolved users may have little commitment to imaging's later success and can easily "second guess" any decision made. Most importantly, without end-user involvement accurate grading cannot occur. As noted in Chapter 1, users must ultimately judge whether the imaging system is "like paper, only better."

Functional Criteria

The basic questions used to organize functional requirements in the Request for Proposal are revisited in the evaluation process. Following are typical considerations, as you weigh vendor solutions:

What documents should be stored as images?

The types of documents stored may include internally generated manuals, reports, microfiche or forms. Although one-time production costs are high, it may be appropriate to offer manuals through CD-ROM technology. Manuals can then be accessed by any workstation with a CD reader, even in remote offices without an imaging server or LAN. Access to computer-generated reports can be provided through COLD (Computer-Output to Laser Disk) to eliminate the physical scanning of these documents. Existing microfiche can also be converted to image automatically with the appropriate technology. This capability may exist in-house or can be performed off-site during backfile conversion. Finally, the vendor's proposal can include software for the design, control and storage of forms. Although not stored as images, *electronic forms* may utilize the same client/server and LAN technology as an imaging system.

Where in the organization will these images be used?

Depending upon whether the desired solution is enterprise-wide, departmental, or a stand-alone workstation, vendor requirements may vary. For departmental or enterprise-wide solutions, the client/server architecture is generally more *scalable* than a mainframe solution. Client/server technology is discussed in more detail in Chapter 6.

Many documents are processed by several staff within one department, or by staff within several departments. Workflow management software can route documents from desk-to-desk automatically, prioritize efforts, and provide status reports on work-in-progress. Some workflow software enables users to develop or modify document flows, using Windows-type icons. Other workflow systems require programming skills, but may be suitable in more static or complex processing environments. The user-friendliness of workflow software should be evaluated by both end-users and IS staff.

Who should be authorized to store, access or modify these images?

Issues of security and integrity, as with other forms of data, also exist with regard to images. Image security can potentially be provided via passwords at the following levels:

♦ system

♦ application

♦ screen

♦ image

After access is authorized, the scope of a user's capabilities should also be defined by the security module. For example, processors authorized to index documents may only be authorized to view documents they have personally scanned. There also should be utilities, such as automated logging of aborted entry attempts, to assist the system administrator in monitoring system access. Most security modules also enforce changing of passwords after a period of time. Finally, if the image serves as a legal record, it is essential that system controls prevent image modification. Legal issues are discussed in detail in Chapter 6.

When will these images be needed?

With imaging, the issue of speed-of-retrieval is more complex than with traditional automation. In traditional automation, data storage is *on-line* or *far-line*. On-line storage usually is some form of magnetic disk, while far-line storage is a magnetic disk or tape, unmounted and stored in a library. By contrast, images can be stored *on-line*, *near-line* or *far-line*. Images stored on-line at the client workstation provide access within five seconds or less. Images are stored near-line if the optical disk is loaded in a jukebox. Image access is provided in 5-60 seconds because the disk must be located robotically and mounted before it is read. As in a traditional tape library, far-line storage can provide access within one minute or several hours, depending on how the library is staffed. The optical disk is retrieved and mounted by an operator, when he or she receives a request.

Consider the time-critical value of the image when evaluating vendor proposals. What is the relative value of access to an image within 24 seconds, 24 minutes or 24 hours? Figure 5.4 presents an example time-value of an image, and the cost of retrieval. The intersection of these two lines (where the cost of retrieval matches the time value

Figure 5.4 Time Value and Retrieval Cost of Data

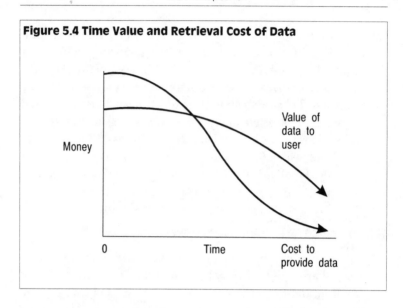

of the image) should represent an upper-limit on any expenditure.

How will the images be accessed and used?

Flexibility of access is the primary advantage of imaging over paper-based or micrographic storage. After a paper document is filed under customer name, for example, it cannot be filed under date received unless a photocopy is made. Much labor and inconsistency in paper-based filing occurs as these duplicate documents proliferate.

Imaging systems with a relational database, in contrast, can provide cross-referenced searching on any data key defined for the document. Such a system might retrieve, for example, all orders over $10,000 received between January and March 1992, from customers with the zip code 32085. Data keys, such as order amount, date received, and zip code, must be defined in advance—but once defined, any combination of keys can be used to locate and sort documents. In contrast, hierarchical databases may limit access to specific search paths and sorting criteria that were defined when the system was installed. This may be entirely sufficient, however, for many types of documents.

For some documents, it is impossible to envision how retrieval needs to occur. With imaging systems used in litigation support, for example, any word or phrase in a document may be used as a search key. A transcript of pretrial statements by a prospective witness is called a deposition, and a law firm may accumulate 100,000 pages or more of depositions in preparation for a trial. Any inconsistency between the statements of witnesses in their depositions and on the witness stand is potentially relevant. As a witness is questioned, another attorney armed with an imaging system can search the deposition transcript and other legal records for inconsistencies. Multimillion-dollar settlements have been attributed to the speed and flexibility of image retrieval.

Related to how a document is retrieved is the question of how it is used once retrieved. Integration with existing applications is a priority with many imaging solutions. Workers traditionally gather documents together on their desk for review, as they perform a CRT inquiry. Imaging replicates this process by combining the application system screen with images of pertinent documents. A loan collector calling a delinquent customer, for example, can view an image of a delinquency notice on their workstation, an image of the signed loan note, and a delinquency screen from the loan-application system.

After a customer's order is scanned and is indexed by customer name, additional customer data may then be available for use as a search key, such as an account number, telephone number or address. Most images are handled simply as one more data element in the application system's existing database.

Technical Concerns

Technical issues or concerns parallel functional criteria, but with an emphasis on the supporting technology. End-users will only confirm that a feature is provided, while IS staff should consider how it is provided and the suitability of a vendor's solution. As shown previously in Figure 3.1, technical issues can be organized to follow the format of end-user questions.

Client/Server Architecture When addressing these questions from a technical perspective, IS staff should consider their organization's approach to emerging issues such as cooperative processing and client/server architecture. Even in large organizations, the era of the mainframe has been superseded by more flexible technologies that separate information processing into its basic components of storage, processing and communication. When these basic components employ *open* or *plug-compatible* technology, they offer smaller, more flexible building blocks within an organization's IS architecture.

In contrast to a monolithic mainframe solution, client/server processing offers an opportunity for incremental investments in technology. And when the foundations for vendor solutions are non-proprietary operating systems such as UNIX, the buyer is freed from passive reliance on a single vendor's software or hardware. Client/server systems, by definition, separate information delivery into its most basic elements: *storage, processing* and *communication.*

Storage of data and images occurs via a data server which may be a microcomputer, minicomputer or even a mainframe. A database residing on the server offers flexible data or image retrieval. If multiple servers exist, data or images are typically stored at the site where they are used most often to speed access and minimize network traffic. In other words, a client/server system stores data and images closer to their users. *Processing* may occur both at the workstation-level or within a centralized application system (residing on a minicomputer or mainframe). Processing occurs at the most logical level. Manipulation of data for departmental reports, for example, occurs on PC workstations within the department. *Communication* occurs via local area networks (LANs) or wide area networks (WANs). The addition of workstations does not generally erode performance of the mainframe application. LANs or WANs can proliferate as they as required, with communication *gateways* provided to mainframe applications.

Just as we have reduced the questions of image handling and processing to the most basic level (Who needs them? When are they needed? Where are they needed?) client/server systems separate IS processing into its most basic elements. The movement toward cooperative processing and client/server architecture is fueled by a need for flexibility that parallels the world-wide trend toward corporate *right-sizing*.

Other Technical Considerations Most IS departments have specific guidelines with regard to evaluation criteria such as modularity of design. Beyond issues of IS architecture, technical considerations in the evaluation include the following:

◆ speed of retrieval

◆ reliability of communications

◆ image integrity/security

◆ scanning, display and printing resolution

◆ ease of use of development tools and integration with existing applications

◆ modularity of design

◆ proprietary versus open systems solutions

◆ technical standards

Figure 5.5 summarizes the current and emerging technical standards in the imaging industry. Some technical issues or standards, such as adherence to a non-proprietary operating system (such as UNIX) may be sufficiently important to be considered mandatory for any vendor solution under evaluation.

Management Criteria

As discussed in Chapter 3 on preparing an RFP, management criteria when evaluating vendors includes the following:

Figure 5.5 Prevailing and Emerging Technical Standards		
	Standards	
Component	**Prevailing**	**Emerging**
Workstation	DOS Microsoft Windows	X, Motif, PM
Server	UNIX	X, POSIX
Electronic Mail	—	x.400
Image Compression	Group 3,4	JPEG
Image-File Format	TIFF	ODA
Database	Relational	Object-oriented
Access Method	SQL	—
Development Languages	C, CBOL	C++
Printer	—	Postscript
Remote Procedures	RPC	OSF/DCE
Optical Disk	—	X3B11
External Devices	SCSI	SCSI II
Network Protocol	Ethernet, Token-ring	FDDI

Source: Nat D. Natraj, "Architectures and Standards: Considerations in Document Imaging Systems," *Document Image Automation*. November–December 1991: 336.

Vendor experience and stability

- ◆ company history
- ◆ financial performance
- ◆ personnel resources
- ◆ installed client base
- ◆ source of capital

Vendor support services

- ◆ project management
- ◆ conversion
- ◆ training
- ◆ customization
- ◆ documentation

Experience and Stability Bud Porter-Roth, a frequent author on the topic of imaging RFPs, recommends that buyers have established criteria for evaluating vendor viability. Potential criteria include company history, financial data for three prior years, the number of people employed, and the current number of installed sites.

Annual and quarterly financial reports are available for publicly traded vendor companies. Closely held vendors may require a non-disclosure statement be signed before "opening their books." For all vendors, a report can be ordered from Dun & Bradstreet to provide an insight into current cash flows. If a vendor is chronically late in payments to suppliers, it will be recorded in their D&B report, and a "cash crunch" is a typical precursor to bankruptcy protection. The vendor's source of capital is also relevant. A start-up firm with a limited history but a viable product may be well-funded by venture capital.

Vendor viability is a difficult question in the selection of an imaging solution because entrepreneurial, undercapitalized companies sometimes offer the most advanced capabilities. This is especially true with regard to research-intensive software or hardware applications such as recognition technologies or artificial intelligence. Many venture companies are formed with an eye toward a profitable merger or sale, once their technology is proven. And while some are sold, others simply fail. An effective compromise may be to assign full financial responsibility to a well-capitalized lead vendor or integrator, who will then bear responsibility for the performance of smaller subcontractors or software houses. A few major integrators, such as GTE/Contel, heavily promote their role as a "financial underwriter" of the small venture firms.

Vendor Support Services A vendor's proposed project plan (deliverables and due dates) are based on its understanding of your needs, as expressed by the functional requirements in your RFP. Even if your requirements are well-defined, changes are likely to occur as you implement an imaging solution. However, the initial project plan provides an essential benchmark, enabling the buyer and the seller to understand and control changes in costs or delivery dates, as needs are modified. Without a well-defined project plan, it is impossible to understand or manage "change orders" later in the project.

The goal of the project-management section of the vendor's proposal is often *agreement in principle* rather than a final schedule of commitments. Review and revision of the management plan and deliverables may be a crucial issue in contract negotiation, and additional revision is likely as the project progresses. The RFP and the vendor's proposal provide a foundation for determining shared responsibilities.

Responsibilities for system installation, documentation and training typically fall to the party responsible for software design. An exception is document *backfile conversion,* which refers to the imaging of existing documents. Depending on the life cycle of a document, no document backfiles may be imaged, or several years' worth of documents may be scanned. Some service firms specialize in backfile conversion. A major backfile conversion demands a large, short-term labor force of conscientious workers. These servicers are experienced in recruiting and managing the temporary workers needed to execute a backfile conversion. Grants are sometimes provided to hire physically disabled workers to assist in backfile conversions. An RFP may be issued to a backfile conversion bureau, or vendors may be asked to subcontract this effort. Considerations relating to backfile conversions are discussed in detail in Chapter 7.

Methods of Validation

While the vendor's proposal can be referenced in the contract, representations should not be taken at face value. Even the most conscientious vendors may misconstrue a question and overstate their capabilities. For that reason, end-users and IS professionals should validate capabilities using the following techniques:

- review of vendor documentation
- demonstrations
- user surveys
- site visits

As previously discussed, your RFP will request that vendors reference their documentation to support all functional claims. As a first step in the verification process, these vendor materials can be reviewed. Methods of grading may include full or partial assignment of points based on the quality of the vendor's approach. Responsibility for validation and grading should be shared by task force members. An internal auditor, for example, would validate system security and control features.

Through your review of documentation, some vendors may be disqualified. For most vendor candidates, however, your evaluation will include a demonstration. Your task force's review of vendor documentation will answer some questions while raising others. A preliminary review of vendor documentation enables your task force to identify and focus on key issues during the demonstrations. These demonstrations should not be the typical "dog and pony show." They should follow the structure of your RFP document. In the demos, the task force will concentrate on grading a vendor's capabilities—documenting the firm's strengths while also identifying weaknesses. It is important not to succumb to vendor hype. Your grading of system features should be pragmatic, based on the specific business goals of the application. "A feature is useless," one imaging user notes, "if it cannot be easily accessed and integrated into a higher-level solution for solving a specific type of problem."[5]

At this stage, it is often most efficient for selected members of the task force to meet individually for one-on-one demonstrations with a vendor's specialists in each application area. These demonstrations should be performed by the vendor's technical staff, not by their marketers. Technical staff are often more candid concerning areas of weakness, and can provide insight into unique strengths. Consider it a subtle warning if you are not allowed to meet informally and privately with a vendor's technical staff.

---

5.   Leonard Yencharis, "What Large-Scale End-Users Want—And What They Get," *Advanced Imaging*, September 1991: 14.

Figure 5.6 User Survey Questionnaire

	Not Applicable	Strongly Disagree				Strongly Agree
1. Software customization services met your expectations.	NA	1	2	3	4	5
2. Document conversion services met your expectations.	NA	1	2	3	4	5
3. Training services met your expectations.	NA	1	2	3	4	5
4. The quality and completeness of documentation met your expectations.	NA	1	2	3	4	5
5. Ongoing support services provided by the vendor met your expectations.	NA	1	2	3	4	5
6. The functionality of the vendor'ssoftware met your expectations.	NA	1	2	3	4	5
7. The performance of the vendor's hardware met your expectations.	NA	1	2	3	4	5

Responses to user surveys can help calibrate the quality of vendor-support services and project-management skills (Figure 5.6). Follow-up calls to respondents can be equally revealing, especially if the vendor obliges by providing a user list that has not been "cherry picked" for the best possible references. A site visit to the vendor's home office is mandatory before any decision is finalized. This visit typically occurs in the context of the final demonstration. Site visits also should be scheduled to customer installations similar in size and scope to the planned system, and ideally a user in the same industry.

Following this chronology (a review of documentation and user surveys followed by a demonstration and site visits) can prevent much wasted effort. If a vendor can be disqualified on the basis of poor user-support or clearly inadequate functionality, your time available for demos and site visits can be concentrated on the most viable candidates.

Cost Projections

Concurrent with the grading effort, one or more task force members should assume responsibility for preparing cost projections. As previously noted, cost is not an evaluation criteria—it is a measure of value, as the price paid in exchange for the functionality and support offered by a vendor. Cost projections should be prepared detailing both initial investment expense and ongoing maintenance expense. These projections are needed for both cost-justification (as discussed in Chapter 4) and for vendor selection.

The initial investment cost includes hardware, software, software customization, installation, conversion, training and communication. Ongoing operating cost includes systems support personnel, hardware maintenance and software maintenance.

Hardware   Hardware encompasses servers, scanners, display workstations, optical storage, printers and communications cabling. For most hardware, the two primary considerations in "price versus performance" are speed-of-retrieval and document resolution. Flatbed scanners are similar in appearance to a small copier, can process four to eight documents per minute, and range in price from $5,000 to $15,000. Automatic sheetfeed scanners with a speed of 40 to 80 pages per minute cost from $50,000 to $100,000 or more. Sheetfeed scanners may be designed to capture one side (simplex) or two sides (duplex). Displays vary in cost based on size, color capabilities and resolution. Cornerstone, for example, is a leading vendor of monitors for imaging systems. Printers also vary in cost based upon speed and resolution.

Optical storage alternatives may vary from dedicated optical drives designed for a single workstation to jukebox configurations that can store up to 2,400 optical disks. The complexity of the jukebox should not be underestimated—its subsystems include drives, media, software interfaces and robotics. "The reliability of the system," according to Tyler Wilkins, director of engineering for Cygnet, "is no better than the weakest link in the subsystem. Stated another way: each subsystem has a failure rate; the sum of the failure rates is the failure rate of the system."[6] If your jukebox supports a mission-critical production application, the maturity of its design (as measured by its release date and installed user base) is an important consideration. Other considerations include the availability of parts and hot-line vendor support. Wilkins warns, "As with other high-tech products, jukeboxes often start shipping before the maturity date has been fully achieved."[7]

Servers may employ proprietary operating systems, or open systems, such as UNIX. In general, vendors with non-proprietary, open systems are preferable because they provide flexi-

6. Tyler Wilkins, "Jukeboxes: Meeting the Reliability Expectations," *Optical Information Systems*, September-October 1990: 252.
7. Ibid., 253.

bility in the choice of hardware. UNIX-based systems, for example, can reside on hardware offered by Data General, DEC, Hewlett-Packard, IBM, NCR or Unisys.

Software An extensive number of software applications is required to manage the scanning, indexing, storage, retrieval and printing of documents. The same hardware may be used for a print server or data server, while specialized software applications provide the unique functionality. In a client/server environment, software controlling the imaging database and centralized optical storage resides on a server. In fact, the "imaging software" itself can be based upon a well-established database package, such as Informix or Oracle, modified to support the storage and retrieval of images. Workflow management software also usually resides on the server.

Workstation software includes PC-based operating systems, such as DOS or OS/2, and usually graphical-user interface (GUI) software that "sits on" the operating system. *Windows* is currently the most widely used GUI application. Image compression and decompression can be performed by software or hardware. Hardware boards for image compression are faster but more expensive than software. A compromise may be to purchase compression boards for frequent users, and compression software for infrequent users.

Software Customization Software customization is a major expense, whether it is performed by your IS department, by an integrator, or by the vendor. As a rule of thumb, first total the hardware and software costs of the proposed imaging solution. For solutions that include workflow routing and application integration, software customization costs are often 30-50 percent of this total. If your RFP asks for software customization by the vendor or an integrator, these costs should be identified separately in the proposals. If you plan to perform this customization internally, it is especially important to appreciate the true cost of this effort. Underestimating customization requirements inevitably leads to project delays, or even worse, a poorly designed solution as short-cuts are taken to minimize this cost.

Installation Support requirements such as UPS (Uninterrupted Power Supply) and cabling are similar to requirements for other comparable automated systems. When determining physical and personnel infrastructure requirements, consider how mission-critical the imaging solution is to your organization. In many instances, an organization's viability directly depends on the imaging system—for processing crucial documents that generate revenue, for servicing customers, or for maintaining records mandated by law.

Your imaging system will be only as strong as the weak link in the chain. For client/server imaging applications residing on a local area network (LAN), for example, access to the images relies on the viability of the LAN. The cost of any mission-critical imaging solution should therefore include infrastructure investments, such as LAN monitoring and maintenance software. LANs are an inherently fragile technology, and an entire LAN can be rendered inoperable by a single faulty wire, hardware card or fuse. Without tools to isolate and diagnose a problem quickly, your operation may be brought to a standstill.

Conversion Your conversion includes the transition from a manual to an image-based process. If imaging is integrated with your application system, conversion also includes your transition to a new release of application software. Your internal conversion costs should be well-documented. A major aspect of conversion may be backfile conversion—scanning a backfile of documents that may be days, months or years old. As previously noted, firms exist that specialize in backfile conversions. Your

RFP should be submitted to these vendors if you want to contract out this effort.

Training Training costs should be identified for both end-users and IS staff. Vendors can provide information on the cost of system operations seminars for IS staff. Your internal trainers, working with the vendor, can estimate the time required to develop application-specific courseware and to deliver end-user training classes. The most cost-effective approach is typically a "train the trainer" process, where IS or vendor professionals teach end-users to serve as trainers within their own departments.

Support Requirements Support requirements will vary, depending on the vendor and your specific application. The most intensive support requirements relate to the *system tuning* needed when you scan and process the massive blocks of data associated with document images. The ebb and flow of these images between servers during your batch and on-line activities may, in some cases, require continual attention. Some systems have as many as five separate monitors that must be actively viewed through a "swivel-chair" technique.

As noted previously, support requirements also vary based on how mission-critical an application is to your organization. Some depth of resource is essential through cross-training of staff. For large imaging installations with a 24-hour-a-day processing cycle, at least three operators are required, with two or three trained IS professionals available via pagers to respond in an emergency. Vendors also offer support contracts that provide their specialists on an "on-call" basis—usually within time frames dictated by your level of need. When considering the cost of these support agreements, also consider the cost if you do not have them—what will be the impact to your business if the imaging system is disabled for an hour, a day or a week?

Vendor Selection

A preliminary benefit/cost ratio can be prepared, based on each vendor's initial proposal. These ratios will be revised as costs are finalized and as capabilities are documented. The cost of a solution typically increases as hidden costs are identified, and capabilities decrease when claimed features cannot be documented. Given the capital constraints in most organizations, there is typically a ceiling on the funds available for an investment in imaging. A maximum level of functionality is also of benefit, beyond which additional features simply reflect more "bells and whistles."

Preparation of cost-projections almost always includes communication with the vendors. Whenever possible, vendors should be given time to respond to questions in writing. The evaluation process frequently leads to some adaptation of the proposals. A different monitor resolution may be chosen, or one high-speed scanner may be substituted for several low-speed units. After a mutual agreement that the requirements have changed, request a revised cost estimate from the vendors. When these are received, new benefit/cost ratios can be calculated. When the final benefit/cost ratios have been prepared, contract negotiations with the highest-ranking vendors can be initiated.

Contract Negotiation

The benefit/cost ratios provide a logical foundation for contract negotiation. Vendors can improve their standing either by *reducing cost* or by *increasing functionality*. Your choice of option you most prefer (lower cost or greater functionality) is an element of your negotiating strategy. From the vendor's perspective, this methodology provides a logical and level playing field. The vendor's sole constraints are their own capabilities and their willingness to bargain on price. The strongest vendors relish this type of objectivity, while the weakest vendors may drop from the race voluntarily.

The most common error made in contract negotiation is to focus on explicit costs while overlooking functional require-ments and service considerations. The goal of contract negotia-tion should be to obtain the greatest level of functionality and service for the cost incurred. The relationship between function-ality and price for three vendors is presented in Figure 5.7. As presented, Vendor A would be the preferred alternative prior to contract negotiation. Based on the level of functionality desired, however, you might choose to negotiate with Vendor B on price, to arrive at an imaging solution that is equally cost-effective with Vendor A (but providing greater functionality at a somewhat higher cost). Alternately, if cost is a primary consideration, you might choose to negotiate with Vendor C on functionality and

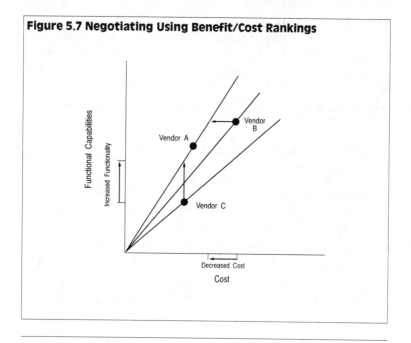

Figure 5.7 Negotiating Using Benefit/Cost Rankings

service, to arrive at a solution somewhat less expensive than Vendor A's (but equally cost-effective).

Effective contract negotiation begins when you understand your organization's needs relating to functionality, training and support—and where you are, and are not, willing to make trade-offs. Armed with this knowledge, and a knowledge of the cost and capabilities of each solution, you are equipped to negotiate a contract that not only controls immediate, explicit costs, but that also supports your future success as you design and implement an imaging system.

6

System Design and Development

"The key to a successful installation is integrating the image system with the corporate information plan, and approaching the planning, design and installation phases in a systematic manner. "

— Connie Moore, Andersen Consulting[1]

Imaging has rightly been called, "a hearty technological soup. " Most imaging solutions are based on a client/server architecture, which intentionally segregates the storage of data from the processing of data through the use of multiple processors, each performing a specialized task. In addition to the complexities typically associated with the diverse components in a client/server system, imaging offers further challenges: unique and often proprietary hardware (such as jukeboxes and scanners); unfamiliar development tools; and increased demands on

---

1.   Connie Moore, "Digital Image Processing in Private Industry," *Optical Information Systems*, May-June 1986: 229.

LAN communications, as images move between optical storage devices and workstations.

As previously noted, a research project in 1990 co-sponsored by Wang Laboratories and the Nolan Norton Institute identified three critical obstacles to successful document imaging:[2]

♦ lack of knowledge concerning imaging

♦ organizational barriers

♦ technological barriers

System design and development is a crucial phase, since all three barriers must be confronted simultaneously. This chapter will discuss technical considerations unique to the development of an imaging solution, and the role of your organization in the design process.

The Client/Server Concept

The goal of client/server systems is to make optimal use of both mainframe and microcomputer technology. In a client/server environment, transaction processing for some batch applications may occur on your mainframe, as it has in the past. But reports or data frequently accessed are stored where they are most use—on a local data *server* supporting multiple *client* workstations. The data server maintains these shared databases and may also be used for some department-level processing tasks. At the workstation level, PCs are used to access and manipulate the data or images.

Servers and clients are joined by a local area network or LAN, as shown by the enterprise-wide imaging system in Figure 6. 1. LANs support two-way communications between all work-

---

2. "Imaging White Paper: Results from the Imaging Working Group," *Optical Information Systems*, May-June, 1990: 146.

stations, and among all workstations and the server. The ability to support two-way communication between multiple processors (both workstations and servers are processors) makes LANs more complex but more flexible than traditional "line drops" from a mainframe to a terminal. A *gateway* in the LAN provides access to the mainframe for data entry or inquiries. The Gartner Group has described traditional computing strategies as *host-centric*. The client/server environment, in contrast, is *PC-centric*. [3]

For your organization's end-users, the value of client/server computing is PC performance combined with mainframe data access. From the perspective of IS, the benefits are equally com-

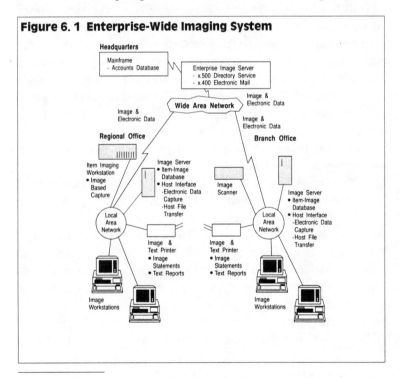

Figure 6. 1 Enterprise-Wide Imaging System

---

3. The Gartner Group, *Conference Papers*, 1992 Annual Symposium, Chicago.

pelling. According to research by Sun Microsystems, client/server computing power (measured as *machine instructions per second*, or MIPS) can be purchased for a tenth of the cost of mainframe computer power.[4] Technical advances have dramatically increased microprocessor performance, while competitive pressures in the open-systems arena (primarily UNIX) have driven down costs. Even when your client/server infrastructure expenses are considered, such as LAN administration and maintenance, the savings are still significant.

Mainframes may always support certain large and transaction-intensive applications. Mainframe resources, however, are not easily scalable—upsizing may depend on a vendor's release of a more powerful system, while downsizing under-utilizes a past investment. The emerging concept of *right-sizing*—precisely matching an organization's human and technical resources to business demand—cannot easily be supported through mainframes. The client/server concept (also called cooperative processing) uses smaller and more specialized building blocks to construct your IS environment. Processing, storage and communication resources are added in small increments, where and when they are required.

The implicit price of client/server computing, however, is an added level of complexity in system integration and performance tuning. "The good news of client/server is that you distribute processing across hardware platforms," according to Robert Boehnlein, vice president of Dome Software in Indianapolis, "and the bad news is that you distribute processing across hardware platforms."[5] From an IS viewpoint, this demands a broader array of technical skills. Supporting a typical client/server environment requires a knowledge of multiple operating systems (DOS, UNIX, and perhaps a mainframe system such as MVS or

---

4.   Sun Microsystems, marketing literature, 1992.
5.   Robert C. Boehnlein, "Integrated Document Database Solutions," Dome
     Software Seminar, Indianapolis, Ind. , April 1992.

VM); graphical-user interfaces that "sit on" these operating systems (Windows, Presentation Manager); database applications to organize data (Oracle, Informix, Sybase, Ingres, Paradox); query languages to formulate requests for data (PC-Focus, SQL); and network-communication expertise. Imaging adds a further layer of complexity—unique software residing both on the client and the server, and specialized hardware such as scanners and jukeboxes.

And the network is emerging as a "system" in its own right, with some theorists suggesting that NIPS (network instructions per second) will replace MIPS (machine instructions per second) as the future benchmark of computing performance. The growing need to tune the entire network, to maximize processor utilization and communication capabilities, should lead to network operating systems that have—embedded within them—advanced LAN monitoring and management tools. The entire amalgamation of technologies (servers, workstations, LANs and the mainframe) would then be a unified, self-monitoring and self-optimizing system.

Given these dramatic changes in technology, the challenge to IS management is articulated by Robert Boehnlein:

> "How do you staff support for developing and maintaining client/server applications? Do you have a team of client specialists, and then a team of server specialists? How do they communicate? What about the middle layer, the network? How will they interact with the team?"[6]

Although client/server computing is a reality, theories abound concerning what the client/server environment will look like in a decade. Some issues, however, can be stated with a fair degree of certainty:

---

6. Ibid.

- IS will move from a host-centric to PC-centric environment.

- While processor costs will decline, the cost of integrating these resources (IS personnel expense) will increase in the near-term as disparate technologies must be merged.

- In the long-term, standardization and open systems (for example, a single version of UNIX), will be forced on vendors by the economics of the marketplace. Business will simply decline to invest in proprietary, stand-alone solutions.

- End-users will emerge as the designers of their own applications. The "glass wall" in IS will be shattered as IS professionals recognize that most computing power (PCs and servers) is in the hands of end-users.

- IS professionals will emerge either as infrastructure specialists (for example, LAN administrators) or as facilitators in the systems-design process.

- Boundaries between IS and the end-user community will become more fluid. End-users who develop technical skills through their use of PCs may transfer into IS. And IS professionals who understand departmental procedures, forms and dataflows may become valued in general management.

Development in a Client/Server Environment

Some system-design issues, such as forms modification or legal requirements, are unique to imaging. More fundamental issues, such as the need for end-user involvement in the design process, are inherent in most client/server applications. It may be meaningful to repeat the questions asked at the beginning of the book:

- What documents should be stored as digitized images?

- Where in the organization will these images be used?

- Who should be authorized to store, retrieve or modify these images?

- When will these images be needed?

- How will they be accessed and used?

Just as these questions reduce the imaging systems-design process to its most basic elements, client/server computing reduces information technology to its most basic elements—separating the storage, retrieval and processing of data or images. The goal of client/server computing is to provide the exact degree and type of technology required, at the precise location it is required. It provides speed, flexibility and responsiveness, at the price of complexity.

For most organizations, imaging represents the "thin edge of the wedge" for client/server computing. Your imaging implementation and system design should teach lessons that can be applied again and again as other emerging technologies—groupware, multimedia and expert systems—are applied in a client/server setting. Your challenge will be to learn these lessons well the first time.

User Involvement

In a recent poll of imaging users, over 64 percent reported that their primary difficulties in installing systems were organizational rather than technical.[7] Given the complex technical challenges inherent with imaging, this statement is noteworthy. The human challenges demand even more attention.

Why would organizations with systems already up-and-running have regrets about the way they were implemented?

7. James E. Breuer, "Where We Stand," *Inform*, September 1992: 68-69.

"They are discovering that basically (the technology) works," says Roger Sullivan of BIS Strategic Decisions, who interpreted the poll results, "but the design was not done well. The introduction was not well-planned. Perhaps it was thrust on the workers without much planning. It's encouraging that it works, but discouraging that they don't involve the line workers. "[8]

It is difficult to believe that any organization would implement an imaging system without inviting the users to participate in designing the system. From every angle of approach—internal communications, training, user acceptance, user awareness of performance and productivity issues, and finally the continuing development (by users) of improved workflow—the participation of frontline workers in systems development is essential. One technique for ensuring end-user involvement is called *Joint Application Design*, or JAD.

JAD Defined

The workflow analysis described in Chapter 2 leads naturally to the development of the imaging applications. The task force engaged to cost-justify and select an imaging solution may also serve as your design group. A phrase coined by IBM to describe a design process involving both end-users and IS staff is called *Joint Application Design*, or JAD. JAD is a refinement of the classical re-engineering process that has been applied in industrial settings for nearly a century. Re-engineering experts (often outside consultants) would study production lines and deliver solutions to increase their efficiency, usually through job simplification and standardization. Both the equipment and workers were regarded as resources to be managed; workers were not usually consulted in this process. In contrast, JAD brings frontline workers into the development process. This occurs for two

---

8.   Interview of Roger Sullivan by Kathleen Hawk, 1992.

reasons: the value of their knowledge and the need for their commitment.

People who do the work inevitably know more about their tasks than anyone else. Even workers who seem to be isolated in highly specialized and fragmented roles can offer unique insights. The commitment and "ownership" of end-users is also necessary. As client/server technology places more processing power and decision-making ability into the hands of frontline workers, their buy-in becomes essential to success. Users who do not understand changes that are occurring can sabotage them by wrong or inappropriate uses of new technology. If some level of agreement with the changes is absent, users can circumvent implementation efforts by resisting training and refusing to abandon old processes, thus "proving" that the new system does not work.

JAD Management

The JAD process typically is managed by an outside consultant, or a professional within in the organization who has expertise in technology and in communication with groups. *Sociotechnology* is one name for this emerging discipline. Understanding how procedures support corporate goals, objectivity and the ability to facilitate discussion are essential skills. A task force leader responsible for earlier phases of an imaging study may also serve as a JAD facilitator.

Work redesign is, by its nature, a departure from business as usual. Internal politics and conservatism retard the process. Managing the process requires strong communication skills, respect for all participants, and the ability to teach a process as well as facilitate it. When successful, Joint Application Design leaves the participants with skills in team-building and group problem-solving that can be used again and again to refine work processes (see Figure 6. 2 for JAD guidelines).

Pilot or Prototype?

In IS nomenclature, *pilot* refers to a one-time, experimental installation, frequently limited in scope, used to establish the feasibility or reliability of a technology. While pilot programs may seem a wise approach to evaluating imaging technology and educating your organization, a number of imaging project managers have concluded that the pilot concept may actually slow

Figure 6.2 Guidelines for Facilitating Joint Application Design (JAD) Sessions.

1. JAD success requires management commitment.

2. Full-time participants must attend the entire session.

3. Make sure you have the right people (including the decision makers) in the session.

4. All participants are equal.

5. JAD preparation is as important as the JAD session itself.

6. Treat all specifications prepared before the session as *proposed.*

7. Off-site and inaccessible means higher productivity.

8. When you start the session on time today, people show up on time tomorrow.

9. Make a good agenda and stick to it.

10. Keep technical jargon to a minimum.

Source: Jane Wood and Denise Silver. *Joint Application Design.* New York: John Wiley & Sons, 1989: 130.

acceptance of imaging within an organization. "Don't call it a pilot," a FileNet executive warned at a recent conference. "Call it a new way to do your work. " Instead of pilot, she suggests relying on *prototyping* with full participation from the user group. "This allows users to envision what you're talking about. If you prototype their forms, it gets them working with you. "[9]

The term *prototype* reflects the inherently flexible and evolving nature of end-user technologies such as imaging. Prototyping is an evolutionary approach to systems development, with an early "rough-cut" system gradually refined through end-user evaluation and suggestions. In one sense, a prototyped system never is completed—it is continually enhanced as users employ it in the workplace and discover new requirements.

Determining Ownership

One of the hottest questions in imaging—and one that directly affects prototype design—is who will ultimately control application development. Or more to the point, where will application development occur? In a user poll by BIS Strategic Decisions at their 1992 Image Systems Conference, nearly fifty percent of the attendees were more interested in "application development tools" than they were in follow-up sales and service from their imaging software vendors.

BIS's Roger Sullivan says this indicates that imaging users want the tools and the know-how to develop their own applications. They want to do it at desktop terminals in their own offices. And they do not want to call the IS department every time they need to make a modification for a short-term project, or a permanent change in the structure of the information flow. "The IS department isn't going to continue to develop applications,"

9. Reporting by Kathleen Hawk at BIS CAP's 1992 Image Systems Conference.

says Sullivan. "Users want a shift to the desktop. If you have to go back to IS every time you make an iteration, you're in trouble. "[10]

This philosophy of end-user involvement has been used to great advantage by USAA, an industry leader in the application of imaging technology. As stated by their CEO, Robert McDermott:

> "We spend an inordinate amount of time keeping users involved from the moment we begin developing any new program. Consequently, when we finally get the product, it's what they need, like and want. They can sit down and say, 'Yes, this feels right.' We use what we call living laboratories, users who test it and kick it and break it during development. When the product is ready, they become the nucleus that goes out and sells it to the first group that's going to use it. They can say, 'You're going to love it, and here's why.'"[11]

When the task force has gone as far as it can in Joint Application Design sessions, the next phase falls on the consultants or internal experts who are familiar with the parameters of the technology. It is their task to design a conceptual prototype that encompasses the issues raised in the JAD sessions. This may include proposed workflow diagrams, approximate system specifications, and a report on the problems and benefits discussed. This report is delivered back to the project team for further discussion. Figure 6.3 presents a *definition document* that summarizes one project's scope, objectives and benefits; Figure 6.4 is a *design document* that shows a screen from the same application.

This begins an iterative process of suggestion and feedback. When the task force has defined a conceptual framework for the new system, it is submitted to the user group for review and

---

10. Interview of Roger Sullivan by Kathleen Hawk, 1992.
11. Thomas Teal, "Service Comes First: An Interview with USAA's Robert F. McDermott," *Harvard Business Review*, September-October, 1991: 127.

Figure 6.3 Definition Document

Project Scope:
This project encompasses the design, development, and implementation of a multi-user document imaging system. Although the scope of this initial project only includes scanning and retrieval of invoices, bills of lading, short bills, manifests, and tally sheets, it is anticipated that this project will follow a strategy that can be expanded to address other document imaging needs of the enterprise.

Project Objective:
The major objectives of the project include:
♦ Provide multiple users on-line access to one year's invoices, bills of lading, short bills, manifests, and tally sheets.
♦ Support one scanning station and three document retrieval stations each with a locally attached laser printer to print documents on demand.
♦ Allow concurrent scanning, indexing, retrieval and printing of documents.
♦ Eliminate the requirement to manually load optical disks.
♦ Provide efficient retrieval of documents by the secondary indexes as well as the primary index (invoice number).

Benefits:
In addition to addressing the specific problems identified above, the benefits of the proposed system include:
♦ Reducing the manual effort to index documents through downloading of more complete indexes and through the use of OCR technology.
♦ Distributing imaging technology directly to the users of the information.
♦ Introducing a single vendor business system strategy (both data and image) that can be expanded and can grow with the requirements of the enterprise.
♦ Providing a strategic foundation to address other document imaging needs—in particular to address and improve customer service levels.

suggestions. When the conceptual design is approved, specific system requirements are defined by the consultants. Finally, when consensus is achieved within the task force, the design is submitted to senior management for approval.

Operational Issues

The design of an imaging system may include hundreds of individual screen layouts for input, storage and retrieval functions (see Figure 6.4), all linked together by complex workflow logic. Although this workflow logic will include procedural opportunities identified by the task force, other operational considerations, including forms design and legal requirements, must also be considered in the systems-development process.

Forms Design

While many imaging systems are acquired to scan and index existing paperwork, opportunities usually exist to redesign documents to facilitate indexing and viewing. For example, Bank of America redesigned over 500 forms to enhance compatibility with an installation of IBM's ImagePlus system. Considerations in the redesign of forms include:

- page size and form
- text-field placement
- font style
- printing standards

Page Size and Layout   With the exception of computer-output to laser disk (COLD) or fiche-to-image conversions, all imaged records begin as paper. A variety of scanners are available to facilitate the imaging of documents. A flatbed scanner looks like a small photocopier, with a glass table where an operator

Figure 6.4 Design Document

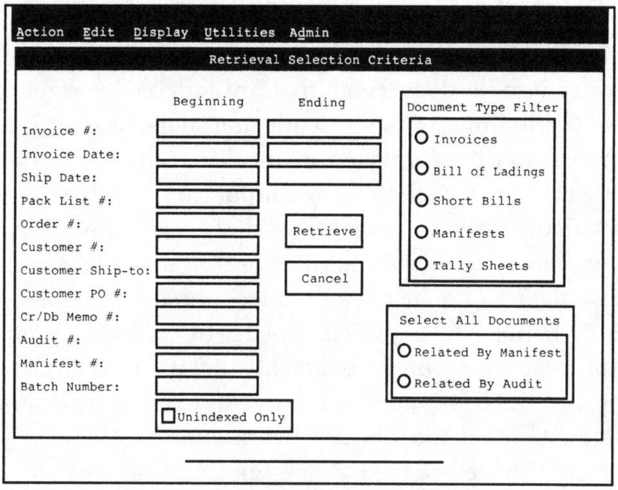

```
┌─────────────────────────────────────────────────────────────┐
│  Action   Edit   Display   Utilities   Admin                  │
│          ┌──────────────────────────────────────────────┐     │
│          │        Retrieval Selection Criteria          │     │
│          │                                              │     │
│          │      Beginning      Ending    ┌─────────────┐│     │
│          │                               │Document Type Filter│
│          │ Invoice #:    [      ] [      ]│ O Invoices  ││     │
│          │ Invoice Date: [      ] [      ]│ O Bill of Ladings│ │
│          │ Ship Date:    [      ] [      ]│ O Short Bills││     │
│          │ Pack List #:  [      ]         │              ││     │
│          │ Order #:      [      ] [Retrieve]│ O Manifests││     │
│          │ Customer #:   [      ]         │              ││     │
│          │ Customer Ship-to:[   ] [Cancel]│ O Tally Sheets││   │
│          │ Customer PO #:[      ]         └─────────────┘│     │
│          │ Cr/Db Memo #: [      ]         ┌─────────────┐│     │
│          │ Audit #:      [      ]         │Select All Documents│
│          │ Manifest #:   [      ]         │ O Related By Manifest│
│          │ Batch Number: [      ]         │ O Related By Audit ││
│          │        [ ]Unindexed Only       └─────────────┘│     │
│          └──────────────────────────────────────────────┘     │
└─────────────────────────────────────────────────────────────┘
```

Title: Retrieval Selection Criteria

Purpose: This screen allows the user to to select the documents to be retrieved and displayed. Users may enter any combination of indexes to specify which documents are to be retrieved. A range of Invoice Numbers, Invoice Dates and/or Ship Dates may be entered to select documents.

The two selection buttons in the "Select All Documents" box can be used to retrieve all related documents with the same Manifest Number or the same Audit Number as the selected documents. These check boxes can only be used when a specific Invoice Number or Pack Slip Number has been entered. These boxes may not be selected if data has been entered in any of the other index selection fields. Both buttons will default to "off."

The five selection buttons in the "Document Type Filter" box can be used to filter out what type of documents are retrieved by the selection criteria. For example, this would allow a user to retrieve only the invoice associated with a specific selection retrieval. All five buttons will default to "on."

places a document face down. Flatbed scanners are suitable for low to medium document volumes.

Flatbed scanners offer the most versatility: an operator can manually handle documents that are tattered or fragile, have staples or bindings, or have an unusual shape or size. Sheetfeed scanners load documents automatically from an input hopper, scan them, and stack them in an output tray. Sheetfeed scanners are available with simplex (one-sided) or duplex (two-sided) scanning ability, and can support higher production volumes. The mechanics of a sheetfeed scanner should be familiar to all users of sophisticated office copiers—it is difficult for these feeders to accommodate documents that are odd-sized, fragile or dog-eared. The volume of documents to be imaged should be considered both in the selection of a scanner and in the design of documents to be scanned. Redesign of documents to permit automatic feeding can reduce the potential for bottlenecks in high-volume settings. Vendors of scanners also can offer recommendations concerning issues such as paper weight and size.

Field Placement and Font Style After scanning, document indexing is the most labor-intensive activity with an imaging system. Optical character recognition, or OCR, allows documents to be indexed automatically, based upon a software program's recognition and matching of a document field against an existing image file or criteria. One section of a loan application, for example, might allow the document to be check-marked as a home-equity loan, installment loan or unsecured loan. Recognition software can then index the application accordingly, if the proper box is marked.

Traditional OCR software seeks to match a character image with a pattern stored in memory. Simple and inexpensive (PC-based applications cost about $500), this technology requires the use of recognizable type fonts, or the placement of notations in predetermined locations on a form. "Omnifont" systems recognize all modern font sets and are widely used to scan and convert hard-copy text into ASCII form. Reader's Digest Association, for example, uses Calera WordScan Plus to convert hard-copy manuscript submissions into electronic files. Although successful with machine-generated type fonts, inexpensive OCR technology has limited ability to recognize handwritten data. Recognition requires the use of simple notations (filling in a blank) in a predetermined location on the form. The Iowa Standardized Test, familiar for years to many students, employed OCR to grade multiple-choice selections.

The weakness in traditional OCR software is that most data submitted on forms is handwritten. And if a department processes several types of forms, they must usually be sorted by clerks that recognize subtle differences between various forms. Successful efforts are underway, however, to develop more advanced recognition technologies. Sophisticated image recognition requires the interpretation of imprecise patterns and incremental learning through trial-and-error. Biological computers—that is, the brains of humans and animals—excel at this type of associative learning; processing is through analog signals, rather than digital signals; identification is through associative searching rather than indexed searching. The weakness of traditional computing in image recognition is due to its precise, digital logic. "Although digital computers can crunch numbers faster than any human, by a factor of perhaps a billion, not even an IBM 3090 can learn to recognize a human face with the speed of a baby," writes OCR researcher David Nordell. "A parakeet can imitate a voice better than a sound digitizer. A worm can learn faster than a PC. "[12]

Beyond the issue of indexing, advancements in recognition technology are having a profound impact on the processing of documents. For $60,000 or more, OCR systems are available that

12. David Nordell, "Inexact Technologies: To Better Mimic the Brain," *Inform*, April 1991: 31.

perform very well on forms with neatly hand-printed block characters. Forms are now being specifically designed for OCR, with special *drop-out inks* that are invisible to the scanner, to separate the form's character boxes from the handwritten characters. The IRS 1040EZ form is the most familiar example of this type of document.

Dramatic savings are achieved with advanced OCR, and with creative solutions to data-entry bottlenecks. Prior to 1989, the Educational Testing Service (ETS) manually processed 3 million financial aid forms submitted by students and parents. Since a typical form required 325 keystrokes, clerks performed almost 1 billion keystrokes annually to log this data. ETS installed a recognition system that achieved high-reliability by training students to draw characters that looked like digits in a hand-held calculator (Figure 6. 5). ETS calls this method "calculator digit recognition" and has achieved very successful results during tests on 1,300 forms. [13]

Recent advancements in OCR technology permit specification of a *confidence level*, stated as a percentage. An organization may require manual review, for example, of images identified with an estimated accuracy of below 95 percent. Accuracy in processing with OCR can also be increased by *cross-verification* of data. A good example of this technique is an income-tax form. Not only is there a complete sum on each page of a form, there are also intermediate results. If any numbers are incorrectly recognized, it can be determined automatically through a comparison of totals and subtotals. Clerks are only obliged to process exceptions.

Concurrent with workflow re-engineering and systems design, opportunities should be sought to facilitate indexing or support processing with OCR technologies. Each OCR vendor

---

13.   Keith S. Reid-Green, "New Digital Recognition Techniques," *Inform*, January 1991: 25-28.

Figure 6.5 Calculator Digit Recognition Set

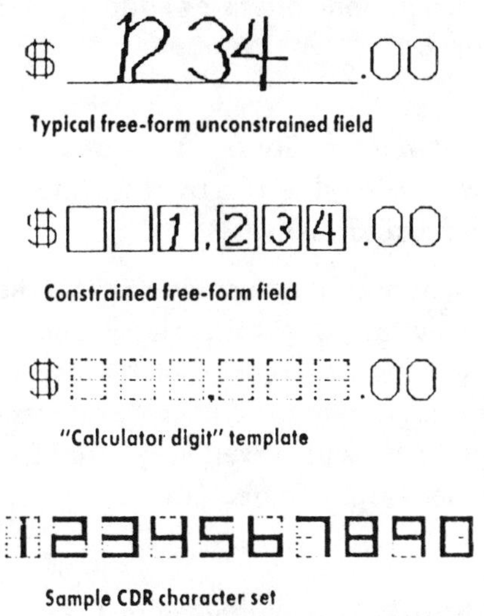

$ _12 34_ .00

Typical free-form unconstrained field

$ ☐ ☐ ☐ [1] . [2] [3] [4] .00

Constrained free-form field

$ ⌐⌐⌐.⌐⌐⌐.00

"Calculator digit" template

1234567890

Sample CDR character set

can provide specific format requirements but, in general, the following issues should be considered:

♦ OCR can be applied to data anywhere on a form, but the location of the data should not vary. Technical specifications given to printers when reproducing a form should be within the defined tolerances of the recognition software.

♦ To assist in the placement of handwritten data, *drop-out ink* can be used that is invisible to the scanner but shows the user where a handwritten entry is needed.

♦ Imaging may be combined with current OCR technologies, such as bar coding, already in use within an organization. For example, a bar-code sticker might be applied to a purchase order to identify a product pulled for shipment.

♦ Some font styles can be read with greater accuracy than others; recognition-software vendors can suggest appropriate fonts.

♦ Handwritten data entered in a defined format will be recognized most easily. As previously discussed, ETS achieved successful results by requiring numeric data in a well-defined format.

Systems that mimic the associative logic of the human brain are called *neural systems*. Neural systems under development will eventually surmount current problems in recognizing handwritten data, and identifying and classifying scanned forms. At present, however, data with a well-structured format and location will require less sophisticated (and less expensive) recognition software.

Printing Standards Documents with a good visual appearance may not appear the same to an optical scanner. "Even though scanners possess more capabilities now than just a few years ago," writes Glenn Kimball, vice president of Areva International, "they will never provide a panacea for all document deficiencies. "[14]

The most significant problem, according to Kimball, is halftone shading. Many documents are shaded everywhere and some documents, such as bank checks, may also be covered with pictorials. The eye perceives the shading on a bank check as having a *numerical average reflectance* of approximately 60 percent (white background paper has a reflectance of 87 percent). An optical scanner, however, will perceive a much lower level of reflectance for this shading. In practical terms, it will not as

14.   Glenn S. Kimball, "Document Design for Electronic Imaging," *IMC Journal*, September-October 1991: 9.

readily discern a contrast between writing on the check and the shaded background.

The reason is that a scanner views a document as a composite of individual *picture elements* or *pixels*. For a scanner with a resolution of 200 pixels-per-inch (also called DPI, or dots-per-inch), each pixel covers an area measuring 1/200 of an inch in diameter. A printing press does not make shaded areas with diluted ink—halftone shading is achieved through the spacing of individual dots of solid ink. A close review of the gray background in a news photograph will reveal this technique. While the eye will visually blend black dots and a white background to perceive gray, the scanner may see the individual dots filling a large portion of the pixel. A scanner with a very high resolution would tend to reproduce individual dots, while a low-resolution scanner would try (probably unsuccessfully) to average the dots to an intermediate level of reflectance. In any event, the result will be background distortion that will limit the efficiency of image compression. As previously discussed, image compression is performed by removing the "white spaces" in a document when the electronic image is stored.

A *threshold* is usually applied in the scanning process, to reduce file size. All pixel reflectance surpassing some designated value or threshold is perceived as "white" and all pixel reflectance less than that value is "black." This can be facilitated by the proper use of ink and paper to maximize background reflectance (or "whiteness"). Backgrounds should be printed with only desaturated (that is, diluted) ink, with a reflectance above 45 percent. The percentage of half-tone dots can be increased to restore the perception of shading.

Issues to consider in developing printing guidelines include:

- ink saturation
- color
- paper opacity
- paper texture
- half-tone screening
- contrast
- reflectance

In general, there is an unfortunate lack of standards or guidelines for printed documents used with imaging systems. As Kimball notes, "Document imaging technology has outstripped the physical document that it is supposed to replace. As long as the physical document must be converted from paper to electronic form, standards must be created or amended to facilitate the conversion process. "[15]

Legal Consideration

Nearly 20 years elapsed before records stored on microfilm were generally accepted as admissible legal evidence. In contrast, the acceptance of magnetic media—disks and tape—as legal records required only 10 years. Significant precedence exists for the acceptance of optical images by courts of law. "Most experts," according to Deborah Kohn, a registered records administrator, "feel there is no reason to delay the application of the new technology until new statutes are approved. "[16]

It is imperative, however, that system developers understand the legal issues relating to image integrity and security.

---

15. Ibid.
16. Deborah Kohn, "The Legal Issues," *Inform*, January 1992: 20-23.

The two possible legal objections to imaging are the *hearsay rule* and the *best-evidence rule*. The hearsay rule requires that evidence come from a primary rather than a secondary source. This potential obstacle is addressed, in the opinion of many attorneys, by the Uniform Commercial Code (UCC) and the Uniform Photographic Copies of Business and Public Records as Evidence Act (UPA). The UCC is a body of law nearly identical from state-to-state, designed to facilitate the conduct of interstate business in the United States. The UCC permits documents to be recorded "in the course of normal business" on "permanent media" if a copy can be reproduced to approximately the original size. While written to permit microfilm storage, the wording is sufficiently general to accommodate optical storage. In addition, the UPA has been adopted by over 30 states and the federal government to address legal issues relating to micrographics and photocopies. Again, attorneys advise that the law's content can be applied without change to imaging technologies.

With regard to the second possible legal objection, the best-evidence rule, state and federal laws exist for successfully overcoming this obstacle. In the 1970s, 26 states and the federal government adopted the "Revised Uniform Rules of Evidence," which classify all computer-stored evidence as "data compilation"—that is, as binary code. The bit map of an image also is stored, of course, as binary code. Some attorneys are therefore advising that destruction of the original document bestows "best-evidence" status, in turn, on the electronically stored imaged.

While a legal foundation for electronic images is now being established, admissibility of a specific image as legal evidence requires absolute proof of image security and integrity. As stated by Danny Proulx, solicitor general of Canada, "Someone will 'jump for joy' at the thought of being able to destroy all the paper once it is recorded on optical disk. It then falls upon the person responsible for systems design to bring that person back to reality by dealing with the archival and legal issues. "[17] Admissibility of an electronic image as legal evidence requires that an organization:

♦ Employ clearly defined and reliable procedures for record entry and audit.

♦ Rely on the electronic images in the course of their daily business.

♦ Prove that the electronic images are secure and cannot be altered.

Procedure manuals should document record-entry routines, and system security should prevent unauthorized entry, indexing or access. An image also should be the "first source" of information, rather than a seldom-used backup, if legal status is a concern. And non-erasable, non-rewritable storage such as WORM (Write-Once, Read Many) may sometimes be preferable to erasable storage for legal reasons.

Most litigation is not resolved in open court but rather in closed administrative hearings in a multitude of federal agencies such as the Social Security Administration, the Environmental Protection Agency and the Internal Revenue Service. Adoption of imaging by government agencies themselves is speeding the acceptance of electronic images as legal records. Three IRS centers currently employ document-imaging technology, and a recent endorsement of optical storage by the Food and Drug Administration has enhanced the legal standing of imaging for

---

17. Danny Proulx, "Optical Disk Technology, Legal Requirements and Archival Retention," IMC Journal, January-February, 1991: 11.

pharmaceutical firms using this technology. In the future, the validity of an image as a legal document will depend more heavily upon the procedures surrounding the technology than upon the technology itself.

7

Installation and Training

"With these technologies ready for legitimate application, we now face the last and possibly the greatest hurdle of all: the user."
— Robert J. Kalthoff, Kalthoff & Associates[1]

Once a system is developed, it must be installed and tested. Documentation will be finalized, and training will be scheduled for both users and IS staff. In addition to the usual complexities of system conversion, imaging adds several dimensions. Existing documents may need to be scanned and indexed in an effort called *backfile conversion*. And user resistance may be strong, if imaging is perceived as a threat rather than as an opportunity.

Brink Trammel of Price Waterhouse cautioned in a presentation at the 1991 AIIM Conference: "Virtually everyone within the organization will be affected (by imaging). It is imperative

---

1.   Robert J. Kalthoff, "The Electronic Image Management Buying Process," *IMC Journal*, July-August, 1990: 14-19.

they understand what is happening so they can solve problems, rather than merely throwing up roadblocks."[2] Communication with the vendor and within the organization can help guarantee success. Trammel advocates a three-step approach to dealing with resistance:

- understand the source of concerns

- address concerns explicitly

- manage perceptions through communication

It is especially important during a difficult implementation to sort out fact from fiction. In one installation plagued by minor, recurring technical problems during its first weeks of production, technical concerns served only as a convenient scapegoat for a mounting backlog of work. Detailed investigation uncovered inadequate training of end-users, unstaffed workstations, poor operational procedures and subtle flaws in the system design. Even if no technical problems had existed (highly unlikely, of course, given the complexity of this technology), the backlog of uncompleted work would have been nearly as great. An impact report can help IS management and end-users sort out crucial problems from those that are merely annoying. This report should identify:

- the nature of the problem

- when the problem began

- the implications of the problem

- a criticality level (on a scale of 1 to 5, for example)

---

2. Brink Trammel, "The Challenge of Acceptance: Imaging's Biggest Hurdle," 1991 AIIM Conference, *Conference Papers*: 344.

♦ who "owns" the problem

♦ when and how the problem was resolved

These reports, reviewed and summarized, can help senior management concentrate their attention on key concerns during the implementation phase. Objectivity, communication and the commitment of all parties to success are all crucial during implementation. If even one of these elements is missing, the entire project may be in jeopardy as minor frustrations wear on the patience of end-users. Technical concerns are inextricably interwoven with human perceptions during implementation. To help lay the groundwork for a successful installation, this chapter addresses technical topics, including site preparation, testing, and documentation, along with issues of training and communication.

Site Preparation

More than one implementation has been delayed by inadequate site preparation. The physical plant must be adequate, of course, for the planned system. The imaging vendor should perform a *site survey* addressing issues such as space constraints, power sources and power quality, network layout and routing, and any ergonomic considerations. "This should take place," cautions integration expert Scott Wallace, "far enough in advance of the scheduled system installation that any problems uncovered by the site survey can be addressed and resolved."[3]

Single workstation installations will have only a minor impact on IS infrastructure. Departmental or enterprise-wide

3. Scott Wallace, *Implementing Electronic Imaging: A Management Perspective* (Arlington, Mass.: Cutter Information Corp., 1991): 69.

solutions, however, will require attention to issues such as electrical consumption, heat generation and space requirements. In general, physical-plant requirements of most imaging components (scanners, OCR devices, printers) are similar to requirements of non-imaging components of the same scale. The jukebox may be an exception. Based on its size and manufacturer, it may demand more power than is typically delivered to the workstation site.

The most significant demands on IS infrastructure, however, will be placed on existing or planned LANs. As Scott Wallace notes, "Image files are much larger than nonimage network traffic . . . and can throttle network throughput."[4] As discussed in Chapter 6, the performance and tuning of client/server devices and their supporting LAN requires more careful analysis than is needed in a mainframe environment. Multiple potential bottlenecks (in the LAN, the server or the client) may reduce response time and processing speed:

- the capacity of RAM storage on the client or server

- the speed of magnetic or optical-disk storage on the client or server

- the processing power of the client or server

- the physical-storage arrangement of images on the optical disk

- the throughput capacity of the network

A number of vendors have introduced "superservers" specifically designed for high-capacity client/server use, and performance modeling tools have become more sophisticated. For the foreseeable future, however, capacity planning and perform-

---

4. Ibid.

ance tuning will remain critical issues in the client/server and imaging environment.

Testing

Responsibility for testing will probably be shared by the vendor and the IS staff, even if the vendor was responsible for development. It is likely that linkages will exist between the imaging system and existing applications, networks and data. The project plan should clearly define responsibility for testing these linkages. At a minimum, IS staff should witness the tests, and it may be preferable for them to develop and execute the tests.

When multiple applications are under development, they should be independently tested and validated, of course, before being tested as an integrated unit. With regard to the network, it may be impossible to effectively model performance because of the numerous and complex performance variables. As the network is tested interactively, the adjustment and upgrading of various components may be needed to achieve desired performance. As discussed previously, clauses in the contract with the vendor or integrator should identify specific performance criteria to be achieved or surpassed. These performance criteria should relate directly to anticipated organizational benefits. If image support is provided to a teleservicing group, for example, retrieval of customer documents should be prompt; if customer callbacks are required due to response times of two or three minutes, the planned purpose of the system is defeated. Vendor payments should be staged, based on the successful completion of each project phase. End-to-end performance testing should be a final condition for full vendor payment. A *user-acceptance test* (UAT) is appropriate to validate that system functionality and performance meets anticipated standards from the end-users' perspective.

Documentation

Responsibility for preparation of documentation was determined during vendor evaluation and contract negotiation. If the vendor or integrator was responsible, the scope and content of documentation should have been clearly defined in the contract. Required documentation will typically include:

♦ system operation manuals

♦ system software manuals

♦ application software manuals

♦ procedures manuals

Conversion

The process of scanning documents into an imaging system is called "document conversion," "backfile conversion" or often simply "conversion." Choices made in conversion planning regarding selected documents, file organization and indexing methods are essentially irrevocable, due to the permanent nature of optical storage and the physical arrangement of stored files on the disks. "Processing decisions made during the system conversion," consultant Donna Bovee emphasizes, "will affect the system's performance and usefulness during its entire life. As a result, practical experience illustrates that a properly planned conversion can result in significant cost savings and improved system capabilities."[5]

---

5. Donna Bovee, "Document Conversion Methodology," *Optical Information Systems*, July-August 1990: 179.

Four major issues should be addressed in planning for conversion:

- What documents should be converted?
- How should the documents be organized and indexed?
- Who should perform the conversion?
- Where and when should the conversion occur?

Identifying Documents to Convert

Identification of documents to be converted was likely performed as needs were defined, or as vendors developed optical-storage requirements. Confirmation of this strategy, however, should occur prior to conversion. The final selection of documents will depend upon:

- records-management practices in your industry
- regulatory requirements
- value of document access
- planned usage of backfile documents

Advice should be sought from records-management staff and your organization's financial and legal advisors. If the organization exists in a regulatory environment, compliance with formal and *de facto* standards should be achieved. Trade groups may publish guidelines concerning record retention, and industry experts may provide assistance.

Beyond the minimum regulatory, financial or legal requirements, consideration should be given to the information value of records. How often are backfile documents needed? How important is this access? Practical solutions may offer the greatest benefit. For example, United Services Automobile Association

(USAA), an insurance company located in San Antonio, Texas, estimated it would take 10 years for a staff of 100 to convert all customer backfiles. Customer records, however, only needed to be retained for seven years. As a logical alternative, they convert customer records only as they are accessed, and estimate that document conversion will be completed in seven years. The conversion of documents should be prioritized based upon practical business criteria: frequency of access, importance of a customer account, or dollar value of the document. "Backfile document conversion can be extremely expensive," cautions Donna Bovee, "and only files active enough to justify such a conversion should be considered candidates."[6] Figure 7.1 presents the retrieval lifecycle for an example document. An explicit understanding of the retrieval patterns for each type of document can prevent your organization from over-investing or under-investing in backfile conversion.

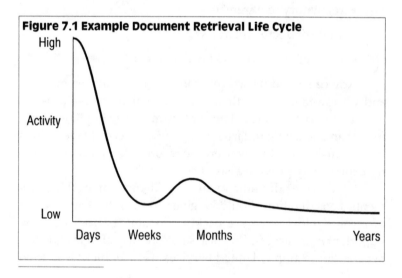

Figure 7.1 Example Document Retrieval Life Cycle

6.   Ibid., 180.

When this analysis is completed, the project manager should prepare the following information for each type of document:

- document category name
- file type and location
- number of documents
- number of pages
- average number of pages per document
- physical characteristics of documents

These figures are essential because the number of documents will be used to estimate indexing workloads, while the number of pages will be used to estimate scanning workloads. Information concerning document location will be needed for logistical purposes, when files are transferred or shipped. And any unusual characteristics (stapled, bound or fragile documents) will be important when planning workloads.

Organize and Index

In general, there are two categories of imaging solutions: *work-in-process systems* and *storage-and-retrieval systems*.

Work-in-process systems support the input, review and analysis of documents. Mortgage-loan origination is an example of work-in-process. With imaging, work-in-process documents are ideally scanned as they enter a department, and are then controlled and monitored by the system. In contrast, storage-and-retrieval systems are used for the archiving of less dynamic documents. Since no processing of documents occurs, files are organized for imaging based on simple criteria, such as frequency of access or date of acquisition.

With work-in-process systems, however, documents are in active use. Conversion must occur while minimizing disruption to work. A phased approach may be taken, with all documents for customers A - D routed to a scanning station on the first day of conversion. During this transition, staff are segregated into two groups: those working with electronic images, and those working with paper. Alternately, work-in-process imaging can be phased-in, beginning with some subset of the process. This piecemeal approach may be suitable for large and complex work-in-process tasks, and will facilitate a gradual testing and validation of system modules.

Related to the organization of records for imaging is the issue of indexing. The most labor-intensive approach is manual indexing of each individual document. An efficient alternative is batch indexing. With batch indexing, an index is assigned to a

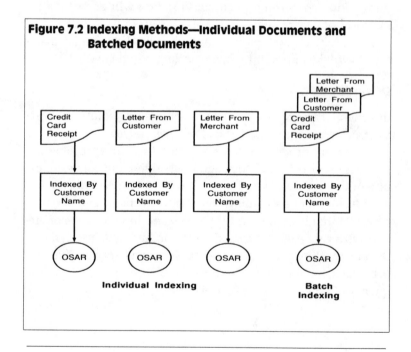

Figure 7.2 Indexing Methods—Individual Documents and Batched Documents

group of documents, followed by scanning of all documents within the batch index (Figure 7.2). This approach is suitable for documents that are filed and retrieved as a unit: all medical records of one patient, all invoices for one customer, all financial statements for one client. When access is needed, all documents are retrieved simultaneously.

The crucial question is the planned usage of the documents. If all documents are not needed concurrently, retrieval of extraneous records will erode system performance. But if documents used concurrently are not stored using a single index, they will probably not be stored as contiguous records on the optical platter. In this worst-case scenario, 12 records for one patient may be stored in an optical jukebox on 12 separate optical platters—and a request for all records will be slowed as various platters are mounted, read and dismounted by the system. Decisions made concerning the indexing, subindexing and organization of files are largely irrevocable, due to the cost and time involved in the conversion effort and the permanent nature of most optical storage (that is, WORM—Write Once, Read Many).

Identify Conversion Agent

An organization without the personnel or expertise to manage document conversion can contract this work to a conversion vendor. Arguments in favor of this approach include well-trained conversion managers and experienced staff. If document conversion occurs gradually following system installation and testing, it is probably feasible to convert documents in-house. If massive backfiles must be converted, use of a *conversion vendor* may be mandatory because of capacity requirements. In such cases, a detailed project plan should be prepared, summarizing the agreement with the conversion vendor. In the contract, performance guarantees should be included regarding throughput and quality.

"The primary benefit of hiring a conversion vendor is the skill, convenience, and time savings that it offers," according to conversion expert Mary Ann O'Connor. "This approach limits the amount of in-house resources required for the project, helps put a ceiling on costs and provides a measure of assurance that conversion objectives will be obtained."[7] Conversion vendors typically quote costs per 1,000 characters converted, rather than on a per-page basis. Some pages may contain 1,000 characters, and others 6,000 characters. A close review of the documents can help in developing an estimate of total costs. Responsibility for related costs such as shipping and handling, document preparation and quality-control review also should be ascertained.

Planning for Conversion

Whether conversion is performed in-house or by a conversion shop, the organization must participate in conversion planning. Issues to consider include:

- ◆ order of conversion, by document category
- ◆ processing procedures and protocols
- ◆ quality-control procedures
- ◆ estimated personnel, equipment and financial resources needed for conversion
- ◆ conversion schedule

Priorities for conversion processing, as noted, can be based on practical considerations. Backfile conversion sometimes occurs before installation, so that images are available as the system

---

7.   Mary Ann O'Connor, "Applications of Optical Technology: Data Capture and Media Conversion," *Document Image Automation*, March-April, 1991: 100.

comes on-line. Alternately, in-process documents may be converted as they are received, with backfile conversion occurring during evening shifts. Whether or not a conversion shop is employed, on-site conversion is advisable if document content is confidential or irreplaceable.

It is essential that procedures be in place to monitor all tasks: document staging, delivery and receipt; document preparation; document movement through the processing facility; and document assembly and return. Scanning and indexing procedures should be defined in writing for in-house or conversion vendors. Logs should be maintained, tracking the movement and location of documents.

Additional equipment resources (scanning and indexing workstations) may be needed for a backfile conversion. This equipment may be bought and then resold after the conversion, or may be leased from a vendor or conversion shop. As discussed previously, the issue of conversion equipment may be addressed during vendor contract negotiation.

Performing Conversion

Whether performed in-house or by a conversion vendor, the conversion effort will include the following tasks:

- request and receipt of documents
- logging and tracking of documents through the process
- preparing documents for image capture
- forwarding documents to operators for scanning and indexing
- recompiling documents and return to storage

If documents are shipped to a vendor for conversion, copies of documents must be retained for safekeeping. This is particularly true when a conversion vendor uses offshore resources. More than one set of documents has been lost in transit. Internal control is equally important (Figure 7.3). The value of protocol and adherence to procedures is articulated by Scott Wallace: "At each conversion station, the flow of documents should be identical. Only one document—the one being converted—is allowed at the conversion station at a time. Staged documents should sit in their batch box, and already converted documents should be placed into a completed box with batch labeling."[8]

Methods of quality control include *display verification, retrieval verification* and *sampling verification*. When a document is scanned, its bit map is placed temporarily in RAM storage at the scanning workstation. This allows an image to be displayed and visually verified, before it is committed to optical storage. An operator will view the image as it would appear to the end-user, and interactively make any necessary adjustments regarding brightness, resolution, rotation or other parameters. The operator will rescan the document if necessary, to achieve proper visual quality. Visual parameters will be optimized for the scanning of an average document, but some documents may be lighter than others, have background shading, or fine detail. Parameters can be changed to ensure the quality of each image. Once the operator presses the "accept" key, the compressed image file is committed to optical storage. Other quality-control procedures should also be performed, including retrieval verification and sampling verification.

Retrieval verification can be performed to confirm that all relevant images—and only relevant images—are being retrieved. The two measures of retrieval effectiveness, *recall* and *precision*, are defined and illustrated in Figure 7.4. *Sampling verification* also should be performed, to confirm that the quality of the stored images is acceptable. Scanned documents must al-

---

8.   Wallace, *Implementing Electronic Imaging*: 78.

Figure 7.3 Murphy's 12 Laws of Document Conversion

1. Expect 20 percent of documents received to require special handling.

2. Needed production equipment will most often fail during the third shift.

3. Take the amount of paper storage and handling room estimated, and then double it.

4. Every hour spent in planning can cut four hours in technical support later.

5. If your calculations of in-house conversion cost aren't within 25 percent of the vendor's prices, something is wrong.

6. Verbal instructions are often misinterpreted, but this will not be discovered until later.

7. The chance of a production machine failing is proportional to how badly it is needed.

8. Complex (that is, expensive) bugs will only be found when production starts rolling.

9. Final acceptance issues will only be discovered after the invoice has been presented.

10. Take all rated equipment speeds and halve them.

11. Backfile conversion becomes a low priority when in-house problems crop up.

12. If an imaging equipment vendor changes a file format standard or specification, the conversion shop will be the last to know.

Source: Gerard Cullen. "Rules of Thumb for Backfile Conversions in Optical Imaging Systems." *Document Image Automation*, May–June 1991: 165.

ways be retained, of course, until quality-control verification is completed.

Process Training

Planning for your training program will likely occur concurrent with your planning for document conversion. A training program should teach skills in work redesign as a foundation for subsequent training in a specific imaging application. *Process training* will therefore precede *technical training*, in a well-thought-out educational program. By participating in process redesign, staff acquire the skills (and hopefully the inclination) to do more of it. They see how change is managed—and note when change is regarded as a positive outcome. Most important, perhaps, they are encouraged to view themselves as change agents. When this perspective is combined with their knowledge of your organization's goals, "smarter" use of technology will inevitably occur.

Figure 7.4 Measures of Effectiveness for Image Retrieval

Terms	Example
$RECALL = \dfrac{\text{number of relevant documents retrieved}}{\text{total number of relevant documents in the database}}$	$\dfrac{9}{14} = 64.3\%$
$PRECISION = \dfrac{\text{number of relevant documents retrieved}}{\text{total number of documents retrieved}}$	$\dfrac{9}{11} = 81.8\%$

Users involved in the planning for new technology take several important steps toward "being smarter." They gain a direct awareness of your organization's "critical success factors." They also begin to understand their role in the organization in both strategic and tactical terms. Strategically, they can see how their performance contributes to the organization's success. On a more task-oriented level, they can see their own role—in terms of workflow and dataflow—in relation to other departments.

At Bank of Boston, the commercial-account group has continued to redesign its workflow after the installation of an imaging-based account-management system. Arthur Flavin, a systems analyst at the bank, says the group—with no assistance from outside consultants—has reworked the account-opening process to reduce it by one day. "My advice to anyone who embarks on an imaging project," he says, "is to work in the department with the people you can get along without the least. Get them involved, pick their brains, teach them how to do process analysis, and you have an ongoing payback."[9]

The redesign process should not happen only once, during the planning process for installation. It should occur repeatedly, as employees adjust their thinking to the potential of paperless workflow. And the system will evolve as the imaging platform becomes the foundation for integrating emerging technologies such as multimedia applications and expert systems. The expertise in identifying opportunities for redesign is at the user level. It is these people who do the work, who live with your organization's limitations, who learn its informal communication networks, and who know how to find things and get things done. The solicitation of ideas and observations from your staff provides more than information. It brings those people into the

9.   Interview of Arthur Flavin by Kathleen Hawk, 1992.

planning process and introduces them to an important concept—that the imaging system can be a catalyst for positive change.

At the same time that workflow analysis and re-engineering are being conducted by the task force, a parallel effort can provide workers with training in teamwork and problem-solving. If attention is focused on "internal customer" issues—the needs of employees up and down the paper-processing assembly line—future imaging users will have an understanding of how their work fits into the larger process. Later, after the imaging system is implemented, frontline staff will have the ability to recognize opportunities to enhance the quality of their work.

The same characteristics that make imaging challenging to cost-justify—the strategic nature of the benefits and the potential for work redesign—mean that the paybacks are cumulative. When another use or benefit is found, it sets the stage for a new path of discovery. Earlier office systems automated *tasks*, such as word processing, while imaging can automate a *process*. As described by one imaging user, Peter Kane of Barclay's Bank, "Taken together, the steps in a business process create a value greater than they do individually. Reviewing a letter from a customer is very common, but when pertinent account information is combined . . . that is very uncommon."[10]

Technical Training

Technical training in your specific imaging application should follow training in change management and process-redesign techniques. For computer-literate staff who are already familiar with the design of the system, this machine training may be

---

10. Peter G. Kane, "Alchemy, Synergy and Other Considerations in Cost-Justifying Imaging Systems," *Inform*, May 1991: 68.

surprisingly brief. One manager reported that a new user following menu-driven instructions located an image within minutes. Others report that departments are "comfortably" working with new systems within several weeks.

But for those unfamiliar with computers, any introduction—however friendly the system—may be intimidating or even overwhelming. Helene Roos, a San Francisco-based consultant specializing in human-resources issues, tells the story of a manager in a state-government agency who assured her that training users on an imaging application was "no problem." On a follow-up visit, her staff interviewed the users and found them alienated and bewildered by the technology. "People who had never used a computer in 30 years of work had been sent to a week-long DOS class," she reports, "because it was available." The class was not only difficult, but meaningless, since the system was Windows-based. "So they came back more confused than ever, and still didn't know how to turn on the machine."[11]

The lesson here is that technical training must be appropriate to the audience, to the technology, and timed so that the users can apply it immediately. The state-government employees needed a simple introduction to Windows, emphasizing the features they would be using, not training in an underlying operating system.

Training should not be limited, of course, to a one-time classroom experience. Staff introduced to a new skill will learn something, but it is doubtful they will absorb every fact they are offered in their first training session. They need additional support in the work environment, which may be provided by a corporate helpline, or "resident experts" specifically trained to provide continuing support to fellow users. One organization

---

11. Kathleen Hawk interview of Helene Roos, 1992.

launched a successful "resident expert" program called HELP, an acronym for "Help Empower Local Professionals," that included participation in an ongoing train-the-trainer program, and invitation to periodic vendor-sponsored seminars.

Implications of Training

The future breadth of imaging's influence is uncertain. However, unimaginable change does not occur in business. Only the change that can be imagined is accepted. So there is a learning curve that will be fast or slow, depending upon how your organization approaches change. These concerns, while they seem abstract, are the basis of decisions that must be made concerning organizational communication and training for imaging. If the system is viewed as a straightforward automation of some paper-moving function, that is all it will be. Clerical staff can easily be trained to move paper with computers rather than from file cabinet to desk. One task force discovered that a novice user, following the instructions on the screen, could find a file in less than five minutes. It is not difficult to operate these systems when they are well-designed.

However, if nothing is taught but how to automate some fragmented part of the workflow, that is all that will be accomplished. And the great potential that exists with imaging—to recreate both the work process and the organization—will be wasted. The link between the re-creation of the organization and the empowerment of staff is echoed by virtually every imaging expert. "The opportunity imaging provides is to recombine fragmented jobs which became fragmented because of paper volume," says Helene Roos, "If the only focus in your training is how to use the buttons, you're missing the point. You have to educate users about the big picture. As they become more sophisticated, they'll begin to find shortcuts in the work process."[12] Decisions made concerning implementation and training will profoundly influence the value of your imaging solution. The broader implications of imaging—the need to evaluate and re-design organizational structures—are presented in the final chapter.

12.   Ibid.

8

Evaluation and Fine-Tuning

"A process of evolution lets the users' own creativeness move the application further along, while increasing the level of acceptance within the organization."
— Peter G. Kane, Barclay's Bank[1]

Common to all important trends in technology—client/server computing, document imaging and object-oriented programming—is a need to align these emerging technologies with your organization's structure and objectives. An extensive effort is always made, of course, to adapt technology to an organization. A large portion of any IS budget, for example, is committed to customizing or rewriting software. Little or no thought, however, is typically given to re-examining an organization and adapting it to a new technology or to a new competitive para-

---

1. Peter G. Kane, "Alchemy, Synergy and Other Considerations in Cost-Justifying an Imaging System," *Inform*, May 1991: 68-69.

digm. A passive attitude toward the interplay of technology, competition and strategy inevitably exposes an organization to more agile challengers.

The phrase "work re-engineering" was coined by Michael Hammer in his widely reprinted article for the *Harvard Business Review*: "Re-engineering Work: Don't Automate, Obliterate."[2] A growing number of practitioners and systems consultants now advocate an even broader view, including all facets of an organization as fair game for reassessment. This inclusive view is conveyed by the "S diagram" in Figure 8.1. Developed by Richard Pascale and Tony Athos, and later adopted by McKinsey & Co. and other consulting firms, this diagram addresses the means by which strategic goals are achieved:[3]

+ hire the right staff (i.e., service-oriented)

+ train staff in the right skills (i.e., leadership)

+ manage them in the style required (i.e., trust, expectation)

+ select values to share with them (i.e., customers first)

+ install the right system (i.e., performance appraisal)

+ improve the structure (i.e., less hierarchy, more communication)

The "S diagram" presented in Figure 8.1 has been modified to show the continual nature of this effort: as one or more of the elements change (in response to internal initiatives or economic forces) other elements are adjusted to achieve an equilibrium or unity of purpose. This approach to corporate change assumes what is becoming known as a "learning organization"—an en-

---

2.   Michael Hammer, "Re-engineering Work: Don't Automate, Obliterate," *Harvard Business Review*, July-August 1990: 104-112.
3.   Charles Hampden-Turner, *Creating Corporate Culture: From Discord to Harmony* (Reading, Mass.: Addison-Wesley, 1992): 89.

Figure 8.1 The "S Diagram"

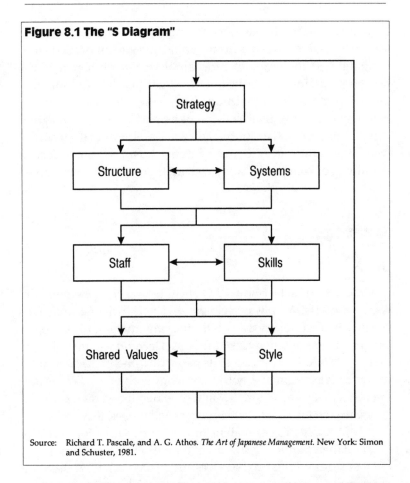

Source: Richard T. Pascale, and A. G. Athos. *The Art of Japanese Management.* New York: Simon and Schuster, 1981.

terprise committed to explicitly and continually re-evaluating its assumptions concerning its culture, its people and its goals.[4]

When IS management assumes a leadership role, not only within IS but within the organization as a whole, imaging may serve as a catalyst to help create this learning organization.

---

4. Peter M. Senge, *The Fifth Discipline: The Art & Practice of The Learning Organization* (New York: Doubleday, 1990).

Imaging advocates in the user community may also provide this leadership. Given the significant capital investment needed with imaging, and the scope of procedural changes inherent in an imaging decision, it is only another small step to conclude that other aspects of the organization be assessed, to ensure that the new whole is more than the sum of its parts. The best imaging installations are not, of course, awkward attempts to graft a new technology onto an old way of doing business, but rather a rebuilding of your organization with tools previously unavailable.

Strategy

USAA

In 1989, Robert McDermott, CEO of United Services Automobile Association (USAA), was recipient of The Gartner Group's Excellence in Technology award. Employing imaging technology, expert systems and networked PCs, McDermott led an initiative that rid USAA of 99 percent of its paper documents. McDermott, however, is a pragmatist, not a technology enthusiast. "We're not interested," says McDermott, "in technology for its own sake . . . only if we can turn it into better services and more satisfying jobs."[5]

USAA began as a small insurance cooperative for the military and their families. Now one of the nation's largest providers of insurance and financial services, much has been written about the pioneering use of imaging technology at USAA. Most readers (and perhaps writers) are unaware that the story actually began many years before the commercial introduction of imaging systems. In 1968, McDermott (then a general in the U.S. Army) was walking through the clerical areas at USAA after working hours.

---

5. Leonard A. Schlesinger and James L. Heskett, "The Service-Driven Service Company," *Harvard Business Review*, September-October 1991: 71-81.

The numerous stacks of paper made him wonder how so many documents could be handled efficiently. To see how the work-flow moved, he discreetly marked an "X" on the top of several stacks of documents. He returned several days later and found most of the marked pages had not been processed or moved. In January 1969, in his first management meeting as CEO of USAA, McDermott challenged USAA to become a paperless company. "By beginning the research early," notes Brink Trammell, a USAA imaging task force member, "USAA would be ready for the technology as soon as the technology was ready for it."[6] The strategic objectives for USAA were service quality and quality of worklife for employees. In one sense, imaging has been instrumental to their strategy, while in another sense, it has only been incidental.

American Airlines

In the 1970s and early 1980s, American Airlines spent millions of dollars building software powerful enough to analyze market conditions and calculate the optimal revenue for its airline seats, flight by flight. The company considered its program the best in the world and for years guarded it jealously. In the mid-1980s, however, American Airlines began to sell this expertise to any-one who wanted it—other airlines, railroads and other potential competitors. Obviously, there was money to be made, but how could the airline destroy its proprietary advantage? The answer, according to Max D. Hopper, IS senior vice president, was not in the software. It was in the expertise of the airline's analysts in using the software. Selling expertise does not diminish the ad-vantage, because in delivering expertise you develop more of it.

American Airlines, which has handbuilt some of the largest and most elaborate information systems in the world, is now

6. Brink Trammel, "Too Little, Too Late: Not at USAA," *Inform*, July-August 1989: 24-26, 50-52.

anticipating buying more hardware and software from third-party vendors. The airline's involvement will be increasingly limited to systems integration, Hopper predicts. If the airline's competitors can buy the same hardware and software, how is their advantage maintained? "We're not in business to build computer systems," states Hopper. "Our job is to lead in applying technology to core business objectives. We don't much worry if the competition also has access to the technology; we think we can be smarter in how we use it."[7]

In his book, *Competitive Strategy*, Michael Porter notes that technology can influence a corporation in three vital ways:[8]

♦ It changes industry structure and, in so doing, alters the rules of competition.

♦ It creates competitive advantage by giving companies new ways to outperform their rivals.

♦ It creates whole new businesses, often from within a company's existing operations.

The competitive solution, in other words, is not in the technology itself, but in the business context in which it exists and is applied.

In the 1990s, imaging may offer some inherent competitive advantage. All emerging technologies, however, eventually mature and become commodities. "Imaging" currently exists as a separate concept only because of its novelty and because it forces decisions about upgrading processing platforms and improving information architectures. By the late 1990s, however, images will simply be acknowledged as another form of business data.

7. Max D. Hopper, "Rattling SABRE—New Ways to Compete on Information," *Harvard Business Review*, May-June, 1990: 118.
8. Michael F. Porter, *Competitive Strategy: Techniques for Analyzing Industries and Competitors*. (New York: Free Press, 1980).

The competitive leaders will not be those who have aggressively purchased this technology, but those who have creatively applied this technology.

Structure and Systems

As cited previously in the Wang study, organizational barriers are one of the three critical obstacles to a successful imaging installation. This is probably inevitable. Traditional organizational structures are paper-flow based. The geography of the workplace was designed to support the physical movement and storage of paper. Organizational structures, in turn, reflect this geography and the inherent limitations of a paper-based system.

One approach to reconciling strategy, structure and systems is popularly known as *information engineering*. Information engineering is a formal analytical approach that integrates strategy, business processes and data into a seamless whole based on future goals rather than past history.[9] "Usually, systems are designed only around an understanding of how the organization works—an understanding of where the information is generated and where it moves," notes Ramiro Valderrama, an information engineer. "A system designed in this approach is often built to serve the organization's current business functions. It assumes the organization is never going to change."[10]

In 1987, First Interstate Bank of California (FICAL) initiated sweeping changes in its retail banking operations. Under the leadership of William Siart, its president, it announced new "Branch of the Future Values" that were to guide all future decisions. In order of priority, these new values were:

---

9. James A. Martin, *Information Engineering: A Trilogy.* (Englewood, N.J.: Prentice-Hall, 1989).
10. Ramiro Valderrama, "Putting the Business Back into Imaging," *Inform*, October 1989: 34.

♦ customer service

♦ sales (revenue enhancement)

♦ expense control (productivity improvement, paper elimination)

♦ loss control (risk management)

This prioritization was significant because, in the words of Gary Sutula, a vice president and regional branch manager: "We took the traditional priorities and stood them on their head. Service, which banks historically neglected, was given first importance."[11] All projects, in progress or proposed, were reassessed based on these priorities. Over 230 automation projects affecting the branches were re-ranked, with additional resources then funneled into the top 10 projects. For one of the major projects, the development of a PC-based prospecting and sales tracking system, interviews were conducted with branch staff and managers statewide. Over a three-month period, more than a hundred interviews were performed, with ideas synthesized into a 350-page design document. The process included the creation of Warnier-Orr diagrams that linked strategic goals, such as service, to specific functions, such as a change of address for a customer. For these functions, improved processes were developed and any required data identified.

Once prepared, this design document was shared for review in a Joint Application Design (JAD) session with representatives from throughout California. As described by Sutula, "Our goal was to integrate function, process and data, based on the four 'Branch of the Future Values'."[12] The result was a system that served to support a future vision, rather than memorialize the past.

---

11. Interview of Gary Sutula by author, 1989.
12. Ibid.

Staff and Skills

Simultaneous with the design of the prospecting and sales system, FICAL management launched a parallel effort to develop the sales skills of branch staff, so that the new system—once installed—could be properly exploited. The system, introduced statewide in 1989, resulted in a dramatic increase in the cross-selling of customer accounts. The experience of First Interstate Bank of California touches on an important issue concerning the role of frontline staff.

There are two diametrically opposed views concerning the role of workers in an image-enhanced environment. Some analysts see a polarization of skills occurring, with fewer but more highly trained managers needed to direct the efforts of frontline workers with increasingly limited authority, experience and responsibility. Educated and highly skilled managers will be involved in the design and monitoring of the system, while frontline staff push the buttons to make it work. Another group of analysts also envision fewer managers, but see frontline workers increasingly empowered by the system—responsible not only for the outputs they produce, but also for the process by which their work is accomplished. In this view, frontline workers must inevitably participate in the design of the system, since they best understand how improvements in the process can be achieved. The first view (which can be called "reductive") is mentioned most frequently by armchair experts. The second view (which can be called "developmental"), is most frequently cited by consultants and IS managers who have borne direct responsibility for the success or failure of an imaging system.

In truth, either approach can become a self-fulfilling prophecy, and even the "reductive" style may generate productivity gains for some imaging applications. The weakness of the reductive approach becomes clear, however, when specific situations are considered. In the 1980s, many retail banks (such as FICAL) spent millions of dollars equipping their branches with PC-based platform automation tools—technology that would simplify and speed the opening and servicing of customer accounts. In theory, these systems might have led to a "de-skilling" of branch staff, since opening new accounts with PCs would simply require following instructions on the screen. In practice, though, even greater skills were required. Training in the use of PCs was the first need, but it soon became apparent that the time saved by the technology could be put to good use. Additional staff training in sales and customer service skills occurred, along with an upgrading of job descriptions from "new account clerks" to "personal bankers." Personal bankers then began to participate in systems design, to exploit the ability of this technology to cross-sell financial services. Although a few banks did use PCs to de-skill their new account clerks, most banks quickly recognized the competitive potential of the technology and turned it to their advantage. In the most profitable retail banks, personal bankers today are often well-paid professionals, highly trained in the selling and servicing of financial products. In the experience of many IS professionals, the "reductive" approach to staffing and training supports the maxim that for every complex problem, there is an answer that is straightforward, simple and wrong.

As the role of frontline workers is changed by automation, the role of the supervisor evolves. This is particularly true with imaging, when line workers have been involved in the design process. When staff are empowered with information about the context of their jobs and are taught team-building skills and change-management models, they not only can revamp workflow, but may in some respects become their own supervisors. And who is supervising these empowered employees? To a large extent, the software will be doing it. The same workflow programs that move images between servers and workstations will also support management tasks, such as prioritizing work, allocating tasks to the appropriate person, and reporting on work-in-process and individual productivity. This does not mean the end of supervisors, but it does indicate an evolution in their reason for being. Previously, supervisors largely became supervisors because of their experience with the work to be done. When workers needed assistance, supervisors would have the answers. Now, with more of the "answers" (including, probably, an entire training sequence) in the software, managers' roles will shift to coaching tasks—such as mentoring, identifying needs for training, and facilitating "skill sharing" among workers. In addition, supervisors may serve as a conduit for information from senior management. The need for retraining is as strong at the supervisory level as it is for frontline workers. "For the supervisor, this is a wonderful opportunity to upgrade the skills set," according to Scott McReady of International Data Corporation. "Instead of having very distinct roles, and all of this social prejudice associated with the role, everyone becomes a dynamic link in the productivity chain. It's a good opportunity to realign roles."[13]

A more challenging concern with regard to supervisors is the issue of restructuring. Imaging may not appear to be good news to employees who learn that their area is targeted for automation. The primary concern, of course, is job security. The need to restructure is implicit when a capital investment in imaging is made for frontline workers. "I would not say imaging is *the* catalyst to eliminating middle-management, but it is one," states Roger Sullivan, a senior vice president at BIS Strategic Decisions. "Imaging is a linchpin of the decision-making process for reorganizing the company and realigning the technical structure. You can blame the technology or give it credit, but the technology is simply incidental. When you empower the end-

13. Interview of Scott McReady by Kathleen Hawk, 1992.

user, you eliminate middle-management. The results are inevitable."[14]

The span of control of supervisors is dictated not by the amount of work processed by staff, for example, but by the number of staff involved in these tasks and the information available concerning their performance. A simple example illustrates this concept. Assume that 100 staff reported to 10 supervisors. Further assume that imaging has increased productivity by 25 percent, and that information concerning individual performance has increased in quality by 50 percent. The same workload should now be processed by 75 staff, reporting to three or four managers.

In this scenario, the only barrier to increased productivity is the willingness to empower frontline workers—by providing them with the technology to solve their own problems, and by trusting they will have the judgment to do so. This broad span of control implies a new role for managers, where they are responsible for coaching and mentoring staff rather than simply supervising them. This also presents an implicit threat to the security of mid-level managers—the very individuals who must support the technology and any organizational change if imaging is to succeed.

In a growing organization, this news is not necessarily negative, since supervisors and middle-management will have two viable options. One is to evolve into a more developmental, mentoring leader with a broader span of control and increased planning responsibilities. Their other option is to bolster the skills on the frontline, by moving into a direct-service position. As the hierarchy becomes flatter, more salary dollars are freed up to recruit and retain the best talent for frontline roles. In practice, some former supervisors have seen little or no reduc-

14.   Interview of Roger Sullivan by Kathleen Hawk, 1992.

tion in their compensation after moving into direct-service positions.

Another cultural issue, as important as restructuring but much less obvious at first glance, is a kind of sensory deprivation. People are accustomed to handling paper, thumbing through it, and measuring their workload and their accomplishment by the movement of stacks of paper. When daily work is shifted to images on a display screen, no matter how big and clear the image, monotony can become a problem. As John Koopman, vice president of PNC Financial Corporation in Pittsburgh, remarked, "We're taking away the warm feeling of being able to touch and feel your workload."[15] There are several approaches to solving this problem. One is to recognize that, in a well-functioning system, there are no natural breaks, no reason to leave the workstation. All the resources can be accessed through the terminal. It becomes essential to schedule breaks, no matter what the press of business. It is also good to provide workers with information about their own performance and achievements, on request. Most workflow software includes management-reporting modules that can give workstation operators access to their own data, historical as well as a record of the current day's accomplishments.

Style and Shared Values

The chief prerequisite for a doctorate in business, or the awarding of a book contract, sometimes appears to be the ability to coin new phrases for an old concept. Although the terminology varies, many analysts classify management styles into two discrete groups:

---

15. Alan J. Ryan, "Soft Issues Challenge Imaging Roll-in," *ComputerWorld*, August 6, 1990: 66.

Top-down	Bottom-up
Theory X	Theory Y
Management-oriented	Leadership-oriented
Transactional	Transformational
Directive	Participatory
Left-brain	Right-brain

A candidate for a teaching position in a Texas school district, when questioned on his personal views concerning Darwinian evolution versus Creationism, is said to have replied: "I can teach it either way." Management science, unlike the physical sciences, offers no definitive answers. An experienced business professional, when asked what management style is appropriate, can honestly answer: "It depends." It depends on the nature of the business, its systems, and its competitors. It also depends on the specific department, or even the individual staff. There appears to be growing recognition of the advantages and disadvantages of both management styles, and a recognition that the best option may simply be to apply the right tool to the right situation.

In his book, *Mind of a Manager, Soul of a Leader*, Craig Hickman calls these two styles management-oriented and leadership-oriented, avoiding a negative connotation for either approach.[16] Hickman suggests that both styles have value and that the best leaders are also good managers—they have mastered the details of day-to-day operations and have thus freed themselves to concentrate on developing their staff. The growing number of books and articles on leadership appears to be an implicit recognition that traditional managerial functions—monitoring, training, performance tracking, exception handling—are now in the domain of the software developers. The evolving

16. Craig Hickman, *Mind of a Manager, Soul of a Leader* (New York: Wiley & Sons, 1990).

ing leader/manager may have two roles, one as a mentor for staff, another as the designer of software tools that monitor their performance. An important opportunity is inherent in this dual role. When the software itself monitors performance, managers can assume a less confrontational and more developmental stance. It is no longer the responsibility of a manager to continually and directly oversee work and report performance deficiencies. It is his or her responsibility to help staff achieve desired performance, once the reporting system has identified a potential concern. These systems are not a substitute for your direct involvement in work performed by staff (and judgment, as always, is needed when interpreting data) but they do allow a change in emphasis toward mentoring and away from "administrivia." The opportunity to assume a developmental, rather than confrontational, role with frontline staff is especially important in the service sector. Research has repeatedly confirmed a correlation between how staff are treated and how, in turn, they treat customers (Figure 8.2).

"In companies that are truly customer-oriented," writes Leonard Schlesinger, a professor in the Service Management Business Group at Harvard Business School, "management has designed (or redesigned) the business to support frontline workers' efforts, and to maximize the value of the impact they create.[17] According to Schlesinger, companies following this approach:

- ◆ Value investments in people as much as investments in machines, and sometimes more.

- ◆ Use technology to support the efforts of men and women on the frontline, not just to monitor or replace them.

- ◆ Make recruitment and training as crucial for clerical workers as for managers and senior executives.

17. Leonard A. Schlesinger and James L. Heskett, *Harvard Business Review*, September-October 1991: 71-81.

Figure 8.2 Correlation between Concern for Employees and Concern for Customers

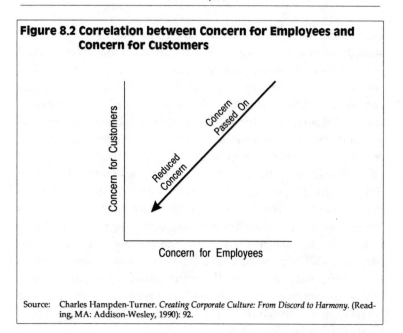

Source: Charles Hampden-Turner. *Creating Corporate Culture: From Discord to Harmony.* (Reading, MA: Addison-Wesley, 1990): 92.

♦ Link compensation to performance for employees at every level, not just at the top.

Companies that do not value frontline workers Schlesinger terms "cycle of failure" companies. "The cycle begins with a set of interlocking assumptions about people, technology, and money derived from old industrial models. Frontline, customer-contact jobs are designed to be as simple and as narrow as possible so they can be filled by almost anyone—idiot-proof jobs." Examples from a variety of companies are cited. Schlesinger's most striking comparison is of two fast-food franchises: Taco Bell and McDonalds.

At Taco Bell, sales grew by over 60 percent between 1988 and 1991, and profits grew by over 25 percent annually (compared with under six percent annually for McDonald's U.S. restaurants). Under Taco Bell's "value strategy," all resources are concentrated on the food, the service and the appearance of the restaurants. These criteria were provided by surveying customers—not by surveying management. After re-examining every aspect of its operations, Taco Bell launched several dramatic changes:

- More than 15 hours of paperwork per week by store managers were eliminated, with an automated system that provides real-time performance data on costs, employees and customer satisfaction. Managers now spend these 15 hours serving customers and coaching frontline workers.

- The span of control was increased from one regional manager for five stores to one regional manager for more than 20 stores.

- The ratio of frontline workers to managerial staff was dramatically increased. These positions were funded by eliminating four layers of middle management. More salary dollars are now spent on service, not supervision.

- A highly selective recruiting process was begun that includes identifying a prospective employee's attitude toward responsibility, teamwork and service.

- Managers are eligible for bonuses that allow them to earn 225 percent of the industry average, based on their restaurant's economic and service performance.

- Managers receive training in communication, team building, coaching and empowerment that they, in turn, pass on to the frontline.

In a fast-food market that was flat or declining, Taco Bell achieved these performance gains while cutting prices in its core menu by more than 25 percent.

The "Learning Organization"

The integration of the various elements of the "S diagram" into a coherent whole is perhaps the greatest challenge that can be faced by management—with or without imaging technology. The ability to create an organization that can objectively view itself and its markets, assess its strengths and weaknesses, and act decisively on this knowledge is an attribute possessed by few corporate leaders. The list of organizations seeking Chapter 11 bankruptcy protection accelerates daily, and even revered, 'blue chip' companies such as IBM have posted historic losses in the 1990s. "Learning disabilities are tragic in children," Peter Senge writes, "but they are fatal in organizations. Few organizations live even half as long as a person—most die before they reach the age of forty."[18]

With imaging installations, a willingness to challenge the validity of long-standing assumptions is especially crucial. Because of the significant capital expenditure required, from $20,000 to $30,000 per workstation, imaging and related client/server technologies can represent a double-or-nothing "bet" for your organization. Before imaging, the cost to dispose of an irrelevant form was simply the time needed to shred it. After imaging, the cost is a rewrite of the software used to scan, index, route and process it—at $100 or more per hour, when contracted out. Before imaging, an initiative to involve frontline staff in process redesign cost only the time and space needed for a task force meeting. After imaging, a fundamental rewrite of the software may be required, if management decides to solicit involvement by frontline staff.

Whether you employ the "S diagram," an information engineering model, or just exercise a healthy skepticism concerning the claims made by imaging vendors, your fundamental concern

---

18. Senge, *The Fifth Discipline.*

is simply how to turn a tool, imaging, into a solution. "This approach (integration with a business plan) puts imaging systems on more solid ground," writes information engineer Ramiro Valderrama. "The systems can be justified not only in terms of dollars, but in terms of how the systems solve business problems. And if much time, money and effort will be expended in the development of a system, should we not clearly keep in mind what the system is for?"[19] In the best imaging installations, the technology has left its fingerprints on the organization, just as the organization has left its fingerprints on the technology.

There is no way to predict what advances in business processes, information services and organizational design will evolve from broad-scale implementation of document imaging. In the early days of the telephone, companies placed their telephones in the mail room, because that is where messages had always arrived. It took time to recognize the benefits of placing a telephone on every desk. Image processing is in a similar historic moment. Culturally, we are so accustomed to the burden of paper we do not see it. The benefits of image processing will be discovered in incremental fashion, as users can envision the changes. Industry is created and is driven by technology and technology, in turn, is driven by our imagination.

---

19.   Valderrama, *Putting the Business Back into Imaging*.

Glossary

Backfile conversion. Refers to the scanning of "backfile" or archived documents onto optical disk. When a significant backfile of documents must be converted, this can be a major operational expense. See also **Conversion vendors**.

Best evidence rule. A legal precedent that when an original document is destroyed, a reproduction of the document (either a photocopy, microfiche, or digitized image), can serve as legal evidence, if reproducible to the size of the original document.

Bit. An abbreviation for "binary digit." A bit is the smallest unit of data, stored in electronic format as either a "0" or "1."

Bit map. Refers to the way an image is stored as digitized data. A "map" replica is created of the document, with data bits used to denote the presence of a black marking. Also called a **Raster pattern**.

CAR. Computer-Assisted Retrieval System. An automated index system linked to a mechanical storage device, that allows requested records on microfiche to be mounted and retrieved. Can also be integrated with an imaging system, to allow micrographics to be retrieved, digitized and displayed electronically.

CCD. Charge-Coupled Device. A drum or belt used to replicate documents, through the attraction of positively charged toner particles that are then imprinted on paper. CCDs are used in laser printers, fax machines and photocopiers.

CD-ROM. Compact-Disc, Read Only Memory. 5.25-inch CDs that store digital data.

Client/server. A computing architecture that separates automation into its most basic components: storage, processing and retrieval. One or more PC servers monitors performs storage and processing functions, while data is retrieved and displayed at PC workstations that may also perform some processing tasks.

COLD. Computer-Output to Laser Disk. Refers to the use of optical disk for storage of computer reports that would otherwise be printed as hard copy or as micrographics. When needed, reports are brought to the display screen for viewing.

COM. Computer-Output Micrographics. Refers to the direct output of computer reports to micrographics. The alternative, printing the reports and then converting them to microfilm or microfiche, is less efficient.

Compression board. A hardware card in an imaging workstation used to compress a scanned image. In layperson's terminology, the compression board "removes the white spaces" from the document, so that only the text or figures are stored. Image compression can also be performed by compression software.

Conversion vendor. A vendor specializing in document back-file conversion. Also may be referred to as a "conversion shop."

Cost avoidance. Refers to the ability of an automated system to prevent the incurring of an operational cost at some future date.

Cost reduction. Refers to the ability of an automated system to result in the immediate elimination of an operational cost.

CRT. Cathode Ray Tube. Also referred to as a "display" or "workstation display." Imaging CRTs are often larger than conventional CRTs (from 17-inch to 19-inch diagonally) to permit full-page viewing of documents, without eyestrain.

DASD. Direct Access Storage Device. Typically refers to the magnetic, hard-disk storage employed by a mainframe computer.

Direct savings. In the context of an imaging system, refers to the elimination of operational costs directly associated with a paper-based system: paper, photocopiers, filing cabinets and the associated clerical support staff.

Display verification. A visual quality control check by an operator that an image has been properly scanned. As necessary, adjustments are made to ensure an appropriate level of brightness and contrast.

DPI. Dots-per-inch. Refers to the level of fine detail provided by a laser printer or an imaging workstation. When in reference to a workstation display, also referred to as pixels, or picture-elements-per-inch.

Drop-out ink. Ink that is invisible to a scanner. Used in forms to limit the amount of optical disk space required to store a form and its data. With drop-out ink, only the data is stored. The image of the form itself is stored only once, as is displayed on the imaging workstation along with the data when viewing is required.

EDC. Electronic Data Capture. The inputting of data electronically, without manual keying. Bar-coding is a common example of EDC.

EDI. Electronic Data Interchange. The direct exchange of data between parties in an electronic format, eliminating the need for manual keying. Direct deposit of payroll is a common example of EDI.

Elective system. A system that personnel can choose (or elect) to use or to not use. To succeed with end-users, it is essential that elective systems be well-designed.

Electronic file cabinet. Refers to simple imaging systems (often single workstation) designed for the retrieval of archived documents. In contrast, more complex imaging systems can support work-in-process: routing documents automatically from workstation to workstation, to facilitate their review and processing.

Electronic forms. Forms that have been designed in an electronic format, with data entry performed directly into the workstation. As with images, electronic forms can be routed intelligently from workstation to workstation, based on forms processing and review requirements. See **Intelligent routing**.

Erasable optical storage. Also known as rewritable storage. This type of optical storage allows data to be erased and new data stored in its place. See **Magneto-Optical storage**.

Far-line storage. Refers to storage of optical disks or other media in a records library, with manual intervention by an operator required to mount the disk. See also **On-line, Near-line** and **Off-line storage.**

Full-text retrieval. An indexing system that allows retrieval of a document based on any word in its text. This indexing is typically performed via OCR scanning of all words in the text. Full-text retrieval systems eliminate the need for key word or subject classification of documents.

GUI. Graphical User Interface. Refers to the use of screen icons on a workstation display to indicate processing tasks (such as a picture of a file cabinet, to convey the filing of a document). Processing tasks are performed through the use of screen pointers and a "mouse." Microsoft's Windows is the most widely-installed GUI.

Hard disk. A term for magnetic disk storage, most often used in reference to a personal computer. Also known as fixed disk, or DASD (when used in reference to magnetic storage for a mainframe).

Imaging. As defined by the Association for Information and Image Management (AIIM), "Imaging is the ability to retrieve, display, process and manage business information in digital form."

Indirect savings. Refers to savings from automation created by gains in end-user productivity. Savings in clerical processing tasks that are automated are direct savings, while time savings for the "consumers" of the data are indirect savings.

Information engineering. A phrase coined by James Martin to describe the integration of data modeling with corporate strategy. See also **Work design.**

Intelligent routing. The programmatic routing of an image from one workstation to another, based upon certain predefined logic. See also **Work-in-process system.**

JAD. Joint Application Design. A technique for involving end-users in the systems design process. See also **Prototyping.**

Jukebox. An optical storage system that uses mechanical arms to locate and mount optical disks. See also **OSAR.**

LAN. Acronym for Local Area Network. A network that supports peer-to-peer communications between any number of multiple processors and peripherals within a defined area (typically within a building). See also **WAN.**

Magneto-optical storage. Also known as M-O. Magneto-Optical storage is a hybrid technology frequently employed for erasable (or rewritable) optical storage. The surface of the optical disk is first heated by a laser, making it susceptible to magnetization. It is then magnetized (that is, the digitized image is stored). After the surface cools, the medium again becomes stable and resistant to magnetization until heated again.

Micrographics. Also referred to as microforms or microfiche, micrographics are miniature transparencies of records.

Multimedia. Refers to the integration of several media, typically image, voice, video and text, in a single application.

Near-line storage. Refers to storage of optical disks in an OSAR or jukebox, allowing a disk to be retrieved mechanically when a stored image is needed. See also **On-line, Far-line** And **Off-line storage.**

Net-tangible-cost model. A technique for evaluating investments in technology that compares the intangible benefits of the investment (such as improved service), to the net difference of the tangible cost of the investment (such as hardware and software) and the tangible benefits (such as personnel reductions).

OCR. Optical Character Recognition. A method for converting an image to data through optical scanning and software designed to interpret handwriting or recognize standard text fonts.

Off-line storage. Storage of data as a paper record. See also **On-line, Near-line** and **Far-line storage.**

On-line storage. Refers to storage of data or an image in internal memory (see **RAM**), on magnetic disk or a dedicated optical drive. See also **Near-line, Far-line** and **Off-line storage.**

Optical storage. Optical storage is not synonymous with document imaging, although optical storage has made imaging financial feasible for many organizations. Also known as OSAR, or Optical Storage and Retrieval system.

OSAR. Optical Storage and Retrieval System. OSAR is typically used in references to jukeboxes, but may also refer to a single drive optical reader.

Pixels. Picture-elements-per-inch. See **DPI.**

PPM. Pages per minute. Used when referring to the speed of laser printers or scanners.

Prerecorded. Optical WORM disks that are produced for data distribution (such as for an on-line encyclopedia). Usually referred to as CD-ROM.

Prototyping. A technique for systems developing involving the development of an early, working model of a system that is continually reviewed and revised based on end-user experience with the system. See also **JAD**.

RAM. Random Access Memory. RAM is the fastest memory storage device, and is used to store images as they are being viewed or manipulated. Also known as "internal memory," RAM memory is a PC chip with capacitors that hold data. Some RAM chips are faster than others. The speed of the RAM is measured by its "refresh rate" or its ability to move data from the memory to the microprocessor. Refresh rate is measured in nanoseconds, or billionths of a second.

Raster pattern. See **Bit map**.

Retrieval verification. A quality control check to verify that an image stored to optical disk can be retrieved.

Rewritable optical storage. Optical storage that, once used, can be erased and rewritten with a new image. Also known as erasable storage.

Right-sizing. Adjusting the size of an organization to current economic conditions. Right-sizing is facilitated by an emerging IS architecture, client/server, that provides smaller "building blocks" to support information needs by separating data processing into its most basic components: storage, processing and retrieval. See **Client/server.**

Scalable. The ability to upsize or downsize an application based on growth or decline in volumes. See **Right-sizing.**

Scanner. Device for creating digitized images of documents. Image capture is performed by reflecting a laser on the document, and recording the light and dark areas as a series of bits.

Sector. A pie-shaped section of an optical disk used to store data or images. Storage of related images on the same sector of a disk speeds retrieval time.

Site survey. A review of an organization's physical plant (utilities and space) to ensure that it can support an imaging system.

Sociotechnology. The emerging discipline studying and improving the interaction of man and machines. Also referred to as "human engineering."

Storage-and-retrieval system. An imaging system used for archival data, also referred to as an electronic file cabinet. In contrast, see **Work-in-process system.**

Systems integrator. An independent consultant specializing in the customization and integration of systems from multiple vendors. In contrast, see **VARs.**

UAT. User Acceptance Test. The final test of acceptance for a new application, following tests of the systems' individual modules performed by the analysts. Also referred to as "end-to-end" testing.

UNIX. An operating system developed in 1969 by Ken Thompson of AT&T Bell Laboratories. UNIX allows a processor to support multiple users and multiple tasks simultaneously. It is used extensively in universities, and increasingly in business. There are several versions of UNIX. IBM's version of UNIX is AIX.

UPS. Uninterrupted Power Supply. Storage batteries used to sustain power to a processor for a defined time period, until a power failure is corrected or power generators brought on-line.

VARs. Value Added Remarketers. Consultants that have contractual relationships with specific vendors, and generate revenue by customizing and installing the vendor's system.

WAN. Wide-Area Network. A network that extends two or more Local Area Networks (see **LAN**), using telephone common carrier lines. A WAN may extend between buildings, or between cities.

Work-in-process system. An imaging system that supports the sequential processing of document images, with tasks that flow from one imaging workstation to another. In contrast, see **Storage-and-retrieval system.**

Work design. Phrase coined in the early 1900s by pioneering industrial engineer, Frederick W. Taylor, to describe the systematic analysis of work tasks and work flows. Also frequently referred to as "work redesign," and more recently, "work reengineering."

WORM. Write-Once, Read Many. Optical storage that can be written to only once. Especially suited for storage of archived documents.

Writable optical storage. Optical disk that can be written to only once. Also known as WORM.

Organizations

Association for Information and
  Image Management (AIIM)
1100 Wayne Avenue, Suite 1100
Silver Spring, MD 20910
301-587-8202

AIIM members receive an industry sourcebook (*AIIM Buying Guide and Membership Directory*), a monthly magazine (*Inform*), and a catalog of books on imaging and other records management technologies. AIIM also sponsors an annual conference and vendor show on imaging technology.

Association of Records Managers and
  Administrators (ARMA International)
4200 Somerset Drive, Suite 215
Prairie Village, KS 66208
913-341-3808

ARMA members receive a monthly newsletter and access to technical publications concerning records management. ARMA

also sponsors an annual conference and vendor show encom-
passing imaging, micrographic and paper-based technologies.

BIS Strategic Decisions
One Longwater Circle
Norwell, MA 02061
617-982-9500

Publishes *Imaging Business Report*, a monthly review of advancements in the imaging industry. Also provides consulting services
and sponsors an annual conference for its imaging clients.

The Gartner Group
56 Top Gallant Road
P.O. Box 10212
Stamford, CT 06904
203-975-6553

Publishes research reports on automation and sponsors several
annual conferences addressing a variety of technologies, including personal computers, networks, client/servers and imaging

Institute of Records Management
4415 West Harrison Street, Suite 200-C
Hillside, IL 60162
708-449-0443

An educational division of Cohasset Associates, a management
consulting firm. Sponsors seminars addressing records management topics that include micrographics and imaging.

The Lafferty Group, Ltd.
Diana House, 4th Floor
33/34 Chiswell Street
London EC1Y 4SE
44-71-782-0590

Publishes research reports for the banking, insurance and bro-
kerage industries on a variety of management topics, including
imaging. Sponsors annual conferences and seminars in key cities
world-wide, and offers proprietary "management briefings" to
individual clients.

Index

About the Publisher

PROBUS PUBLISHING COMPANY

Probus Publishing Company fills the informational needs of today's
business professional by publishing authoritative, quality books on
timely and relevant topics, including:

- Investing
- Futures/Options Trading
- Banking
- Finance
- Marketing and Sales
- Manufacturing and Project Management
- Personal Finance, Real Estate, Insurance and Estate Planning
- Entrepreneurship
- Management

Probus books are available at quantity discounts when purchased for
business, educational or sales promotional use. For more information,
please call the Director, Corporate/Institutional Sales at 1-800-998-4644,
or write:

Director, Corporate/Institutional Sales
Probus Publishing Company
1925 N. Clybourn Avenue
Chicago, Illinois 60614
FAX (312) 868-6250